T0392080

Constructive Axiomatics for Spacetime Physics

Constructive Axiomatics for Spacetime Physics

EMILY ADLAM
NIELS LINNEMANN
JAMES READ

Great Clarendon Street, Oxford, OX2 6DP,
United Kingdom

Oxford University Press is a department of the University of Oxford.
It furthers the University's objective of excellence in research, scholarship,
and education by publishing worldwide. Oxford is a registered trade mark of
Oxford University Press in the UK and in certain other countries.

Published in the United States of America by Oxford University Press
198 Madison Avenue, New York, NY 10016, United States of America

British Library Cataloguing in Publication Data
Data available

Library of Congress Control Number: 2025930996

ISBN 9780198922377

DOI: 10.1093/9780198922391.001.0001

The distinction between axiom and definition leads to the separation of empirical content from arbitrary concept formation, and the derivations of particular statements clearly reveal the empirical and the logical components of every assertion. The rigor of demonstration leads at the same time to exhaustive formulations of the assumptions and automatically excludes the formation of vague concepts.

— Hans Reichenbach, Stuttgart, 1924 [1969, p. xiii]

Acknowledgements

We are very grateful to Jeremy Butterfield, Adam Caulton, Erik Curiel, Richard Dawid, Juliusz Doboszewski, Dominic Dold, Sam Fletcher, Gregor Gajic, Clément-Olivier Hongler, Dennis Lehmkuhl, Joanna Luc, Eleanor March, Niels Martens, Vladimir Matveev, Ruward Mulder, Brian Pitts, Oliver Pooley, Dominic Ryder, Erhard Scholz, Norman Sieroka, Chris Smeenk, Noah Stemeroff, Will Wolf, to the attendees of the Bonn work in progress group, to the Harvard Black Hole Initiative foundations group, to the Stockholm theoretical philosophy group, and to audiences at the Oxford Philosophy of Physics seminar and at the 2024 *New Directions* conference in Viterbo, for valuable feedback.

Our sincere thanks to Peter Momtchiloff, Jamie Mortimer, and Sandhiya Babu, as well as other members of Oxford University Press and the Integra typesetting team, for their guidance over the course of the process of publication of this book. Our thanks also to Franciszek Cudek for preparing the index.

The diagrams in Chapter 1 have been reproduced from those in the original Ehlers–Pirani–Schild article, the copyright for which is held by Oxford University Press and the Royal Irish Academy of Sciences; we are grateful to Andrzej Krasinski for providing us with guidance on the copyright issues related to these images.

Contents

Introduction

What does it take to really understand a theory—particularly a highly mathematized and abstract theory of the kind encountered in modern theoretical physics? The programme of *constructive axiomatics*, promulgated by the great Hans Reichenbach one century ago in his 1924 book *Axiomatik der relativistischen Raum-Zeit-Lehre* (translated into English by Maria Reichenbach in 1969 as *Axiomatisation of the Theory of Relativity*), offers the following answer to this question: we must begin with elementary axioms with direct empirical content, and proceed to demonstrate how the entire edifice of the theory in question can be built up from those axioms. In this way, the thought goes, the theory can be set on a solid conceptual and empirical footing.

These constructive axiomatizations of physical theories have not been lost entirely to history. Much more recently, Carrier writes:[1]

> In a constructive axiomatization only those statements are accepted as fundamental that are immediately amenable to experimental control. A deductive axiomatization, by contrast, confers fundamental status to more abstract statements (such as variational principles). (Carrier, 1990, p. 370)

We'll return later to the details of the content of this passage. For now, note that these 'constructive' axiomatizations, using axioms selected specifically for emprical transparency, are distinct both in motivation and in content from the more common 'deductive' axiomatizations which use axioms selected for (say) mathematical simplicity; constructive axiomatizations thereby offer a number of novel philosophical advantages.[2] For example, constructive axiomatizations of physical theories allow one better to appreciate and distinguish between the empirical and conventional inputs to a theory; moreover, from the point of view of theory construction, they also offer a systematic means by which a theory might be extended and generalized. Although Reichenbach himself, of course, was writing in the heyday of logical positivism—and, indeed, the attraction of constructive axiomatics to a positivist agenda should

[1] Although in this book we take inspiration from Carrier and use some of his terminology, neither our central reconstruction of the Ehlers–Pirani–Schild constructive axiomatization (following Carnap) nor our ultimate case for the usefulness of such a programme is related to Carrier (who ultimately draws a rather pessimistic picture of constructive axiomatics).

[2] To be clear, the constructive/deductive distinction as presented above is due to Reichenbach—both forms of axiomatization are deductive in the logical sense.

Constructive Axiomatics for Spacetime Physics. Emily Adlam, Niels Linnemann, and James Read, Oxford University Press. DOI: 10.1093/9780198922391.003.0001

be clear—constructive axiomatics can continue to offer such advantages even when prised away from that context. Especially since the programme has been sorely neglected in recent decades (with a few notable exceptions which we will consider later in this book), its centenary therefore invites renewed reflection on what it has achieved, and what it can offer going forward.

Reichenbach himself offered a particular approach to constructive axiomatics: he sought to build up the structure of general relativity from the paths of light rays alone. But Hermann Weyl pointed out to Reichenbach (who remained stubborn) that this is impossible; one needs the motions of both freely-falling particles and light rays in order to arrive at the full spacetime structure of general relativity. Still, Weyl's own approach remained programmatic, and the constructive axiomatization of general relativity was only (the consensus goes) realized in full glory by Jürgen Ehlers, Felix Pirani, and Alfred Schild (henceforth EPS) in a very influential paper published in 1972 (Ehlers et al. (2012)).

In this book, we focus primarily on this EPS approach, which remains the most fully developed example of a constructive axiomatic approach to a spacetime theory. We aim to familiarize the reader with constructive axiomatics, to scrutinize the programme's merits, and to demonstrate how the programme can be used to arrive at a better understanding of classical spacetimes theories and potentially even issues in quantum gravity. We hope this book will help to open up constructive axiomatics to new audiences, both in philosophy and in physics, as well as to help to open up a rich and fresh array of new topics for pursuit in the foundations of spacetime theories.

Here is our plan. We begin in **Chapter 1** with a walkthrough to the original EPS constructive axiomatization of general relativity: examining every step, filling in loopholes and proofs, and pointing out unstated assumptions. The latter is particularly important, because only once the assumptions are made explicit it is possible to contemplate removing some assumptions, thus demonstrating that the empirical axioms from which EPS begin are compatible with a richer class of spacetime theories than just general relativity.

In **Chapter 2** we offer a more general discussion of constructive axiomatics for spacetime theories, carefully systematizing the post-EPS literature and the other approaches which have been taken up since the 1970s. We focus on some important discussions of the EPS paper in the subsequent literature, particularly the claim of Sklar (1977) that the whole approach is circular, and the attempt of Perlick (1987) to provide explicit 'clock criteria' to plug some major holes in the approach. We also show how the approach relates to the philosophy of major figures such as Carnap (1967) and, more recently, Brown (2005).

We then move to considering possibilities for future research in constructive axiomatics. In **Chapter 3**, we demonstrate explicitly that if one begins not with particles and light rays as in the original EPS derivation, but instead with branching structures as appropriate for quantum mechanics, one arrives at spacetime theories in which the very structure of space and time branches. This is not only interesting itself, but also presents important lessons for quantum gravity, for it suggests that those wishing to reconcile spacetime with quantum mechanics should be thinking seriously about branching spacetimes, perhaps as represented by non-Hausdorff manifolds. We also show that setting up constructive axiomatics for a branching spacetime requires choices to be made regarding how to think about the identity of spacetime points across branches: a topic which has intriguing links to the hole argument.

Finally, in **Chapter 4** we demonstrate that appropriate modifications of the EPS axioms will yield, not general relativity, but curved spacetime non-relativistic theories, called Newton–Cartan theories (i.e., rationally reconstructing, the predecessor theories to general relativity, rather than the successor theories). This provides a better understanding not only of the constituents of relativistic theories and of the empirical content of those theories, but also of non-relativistic theories; thereby, we learn more about the structure of all possible spacetime theories, in the spirit of a recent exhortation for philosophers of physics to explore the 'space of spacetime theories' (Lehmkuhl et al., 2016).

1

Walkthrough to the Ehlers–Pirani–Schild Axiomatization

Introduction

The 1972 Ehlers-Pirani-Schild (EPS) axiomatization of (the kinematics of) general relativity (GR) is arguably one of the most well-known and highly regarded rational reconstructions in all of physics.[1] Characteristically, the scheme is both (i) constructively axiomatic in the sense of Reichenbach,[2] i.e. builds on a basis of empirically supposedly indubitable posits, and (ii) constructivist in the sense of Carnap,[3] i.e. proceeds semantically in a linear, non-circular fashion. It is, therefore, an ideal case study for the motivations, merits, and conceptual issues involved in both kinds of such rational reconstructions. More specifically, the EPS axiomatization is of profound significance both to the epistemology of spacetime theories (for the result seems to show that we can recover the spatiotemporal structure of the world from the input of empirically grounded axioms, or so the ambition goes) and the metaphysics of spacetime theories (for one might use the result to even argue that spacetime structure is ontologically reduced to the structures invoked in empirically grounded axioms).[4]

This chapter provides a detailed walkthrough of the original work by EPS. Such a detailed presentation of the EPS scheme is prompted by the recognition that the original presentation, for all its beauty, leaves out important motivational remarks, proofs, and clarifications; at the same time, the overall rational

[1] For well-known general work on rational reconstructions, see e.g. Laudan (1978).

[2] See Reichenbach (1969 [1924]). We discuss further Reichenbach in Chapter 2.

[3] We discuss further Carnap's notion of 'constructivist' on p. 10.

[4] It is worth noting that—although their work is clearly an example of constructive axiomatics—EPS do not themselves refer to their work as a case of such; moreover, although explicitly indebted to Weyl, they make no explicit reference to Reichenbach as the progenitor of the tradition. Indeed, even in their other (published) works we have failed to identify any explicit mention of Reichenbach in this context. It strikes us as implausible that EPS could have been ignorant of Reichenbach's work on these issues; however, a full historical investigation into whether any of the three authors made reference to Reichenbach on constructive axiomatics in their unpublished writings or letters will have to wait for another occasion. (Our thanks to Dennis Lehmkuhl and Erhard Scholz for some helpful discussions here.)

Constructive Axiomatics for Spacetime Physics. Emily Adlam, Niels Linnemann, and James Read, Oxford University Press. © Emily Adlam, Niels Linnemann, and James Read (2025). DOI: 10.1093/9780198922391.003.0002

reconstruction can only properly be assessed if all specific steps are clearly laid out and often-ignored ambiguities are rendered explicit. This chapter also provides the actual analysis of the status of the EPS scheme *qua* rational reconstruction.

To facilitate comparison of our presentation to that of the original EPS paper, we mainly follow the structure of EPS' original presentation; in particular, the sequential sections of this chapter concern the respective sections in the original EPS paper on the background assumptions (preamble), construction of differential topology, conformal structure, projective structure, Weyl structure, and finally Lorentzian structure. (We also refer to EPS' figures in their original numbering.)[5] Among other things, implicit assumptions will be made explicit, the form of constructive and constructivist reasoning clarified, and missing proofs and alternative pathways upon modifying certain assumptions illustrated. The final section concerns immediate evaluative questions regarding the EPS scheme in light of this walkthrough.

1.1 Setup

The structures from which EPS begin their constructive axiomatization of GR are the following:

Definition (Set of events). *A point set* $M = \{p, q, \ldots\}$ *is called a* set of events.

Definition (Light rays and particles). *The elements of* $\mathcal{L} = \{L, N, \ldots\}$ *and* $\mathcal{P} = \{P, Q, \ldots\}$*, where* $L, N, \ldots, P, Q, \ldots$ *are all subsets of* M*, are respectively called* light rays *and* particles.

Some remarks here are in order. First: M is a bare set, not yet structured as (say) a differentiable manifold. Second: the same holds for $\mathcal{L}$ and $\mathcal{P}$. In a sense then, it is too early to call $\mathcal{L}$ and $\mathcal{P}$ 'light rays' and 'particles'. Moreover, we just said earlier that the EPS approach to spacetime is constructive in the sense of building on empirically indubitable posits. Admittedly, though, carving up our experiences in terms of events is already quite a theory-laden precursor to a four-dimensional way of thinking.

The decisive structures to be defined consecutively and determined empirically in setting up the kinematical structure of GR as sketched above can be characterised loosely in the terminology of EPS as follows (Ehlers et al., 2012 [1972], p. 65):

[5] All figures have been redrawn for this book.

1. *Differential (topological) structure*: Allowing for the definition of (smoothly) differentiable structures (such as functions) on M.[6]
2. *Conformal structure:* A field of infinitesimal null cones defined over M.
3. *Projective structure:* A family of curves, called *geodesics*, whose members behave in (the second-order) infinitesimal neighbourhood of each point $p \in M$ as straight lines in projective four-space.[7]
4. *Affine structure:* Projective structure, with the addition that the preferred curves carry preferred *affine parameters* such that, infinitesimally, there is an affine geometry around each point $p \in M$.
5. *Metric structure:* Assigns to any pair of adjacent points of M a number called its *separation*.[8]

The characterisations of the above structures (2)-(4) are of infinitesimal kind, in contrast with the (arguably) more familiar definitions of these structures in an algebraic fashion. Notably, the former kind of definitions are more suited to the constructivist project, as on the infinitesimal versions the weaker structures do not have to be defined by way of recourse to stronger structures (which may easily run counter to the principle of Carnapian semantic linearity, discussed further below), which, however, is often the case for the more familiar algebraic versions. We will now show explicitly the equivalence of the infinitesimal and the algebraic notions.

First, a conformal structure $\mathcal{C}$ is usually defined algebraically by presupposing a collection of Lorentzian metrics $g, g', \ldots$ on a differentiable manifold M which satisfy (Matveev and Scholz, 2020, §2)

$$g \overset{\mathcal{C}}{\equiv} g' \Leftrightarrow \exists \text{ a function } f \text{ on } M \text{ s.t. } g' = e^f g.$$

[6] Although by 'differential structure' EPS mean that which allows for the definition of *smooth* structures (e.g. functions) on M, in principle one can countenance weaker notions of differential structure, allowing for the definition of only C^k (for finite k) differentiable structures (e.g. functions) on M. (Indeed, EPS do this at certain points, when e.g. they write 'C^3 seems sufficient for our purposes' (Ehlers et al., 2012 [1972], p. 70) in the context of axiom D_1, or define the g function in axiom L_1 as only C^2 differentiable.) By buying into such smoothness assumptions, one has already imposed certain constraints on the resulting kinematical structures derivable from the ensuing axiomatization (e.g. low regularity versions of general relativity (see e.g. Sämann and Steinbauer (2018))—would be excluded).

[7] Sometimes, authors use the terminology of 'geodesics' versus 'pregeodesics' for, respectively, (up to affine transformations) parameterised versus unparameterised geodesics. In this book, we will use only the latter set of terminology, and unless otherwise stated a 'geodesic' does not come equipped with a determinate parameterization at all.

[8] This is EPS' language. If one prefers (perhaps because one regards the notion of 'adjacent points' as being unclear), one can read this as: 'Assigns to any pair of points in the infinitesimal neighbourhood of each $p \in M$ a number called its *separation*.'

The equivalence between this algebraic definition of conformal structure and the infinitesimal version used by EPS can be seen as follows. Consider a cone around the origin of the tangent space at point p: it is given by $-T^2/a + X^2/b + Y^2/c + Z^2/d = 0$ where we take $\{T, X, Y, Z\}$ to be the canonical coordinates in the tangent space T_pM spanned by $\partial/\partial t, \partial/\partial x, \partial/\partial y, \partial/\partial z$ with $\{t, x, y, z\}$ a local coordinate chart on M around p.

Now, the cone equation can be rewritten directly as $g_{\mu\nu}X^\mu X^\nu = 0$ with $g_{00} = -C/a, g_{11} = C/b, g_{22} = C/c, g_{33} = C/d$ with C a constant *at each point*. Due to the diagonal form in the presupposed coordinates, it should be clear then that $g_{\mu\nu}$ is a Lorentzian metric, determined up to a constant factor C at each point p—this all-together then captures conformal structure. At the same time, there is a one-to-one mapping of the infinitesimal cone structure onto the manifold, meaning that the cone structure in tangent space can indeed be taken to define an infinitesimal cone in a small neighbourhood around p (this follows from the properties of the tangent space itself). So both conformal structure as well as the presence of infinitesimal cones at each p are equivalent.

This, arguably, is how an infinitesimal cone structure at a point is associated with one and only one class of conformal metric structure at a point. Note, however, that should one wish to extend this cone in the manifold to a neighbourhood of $p \in M$, then this will require recourse to e.g. the exponential map construction, which in turn requires an affine connection, which will not be available prior to the construction of some affine connection in the EPS scheme. EPS in fact do not need to require this construction, but to the extent that they might wish to avail themselves of it, we propose on their behalf to take claims of constructing cone structure in a neighbourhood of any $p \in M$ in the following to be also implicit claims about the existence of a suitable isomorphic map between manifold neighbourhood and tangent space. Even if this posit implied—it does not to our knowledge—that there had to be some notion of exponential map and, with it, of affine connection, it can be left empirically unspecified by the conformal structure at this stage which one affine connection and *a fortiori* which exponential map exactly.[9]

Second, an affine connection in the usual algebraic sense engenders a notion of parallel transport, including the notion of a geodesic (a curve along which

[9] It could of course be that the EPS scheme becomes empirically/methodologically inadequate because the presupposition of the existence of some affine connection (its exact nature being unspecified) turns out to be invalid. However, such a possibility only implies a limitation of the EPS scheme (not, as one might think here, a circularity); the point is analogous to one raised in a critique towards (and consequently rebutted by) Coleman and Korté; see Chapter 2.

its own tangent is parallel transported). A converse approach is also possible: given a notion of parallel transport $P[\gamma]^t_s : T_{\gamma(s)}M \to T_{\gamma(t)}M$ with $P(\gamma)^s_s = \text{Id}$, $P(\gamma)^t_u \circ P(\gamma)^u_s = P(\gamma)^t_s$ and a smooth dependence of P on the curve γ and points s, t, an affine connection can be defined in terms of parallel propagation via the relation[10]

$$\nabla_X V = \lim_{t \to 0} \frac{P(\gamma)^0_t V_{\gamma(t)} - V_{\gamma(0)}}{t} = \frac{d}{dt} P(\gamma)^0_t V_{\gamma(t)} |_{t=0}.$$

Third, a projective structure $\mathcal{P}$ is usually defined in the algebraic fashion by piggybacking on a class of affine connections $\Gamma, \Gamma', \ldots$ on the manifold M as follows:

$$\Gamma \overset{\mathcal{P}}{\equiv} \Gamma' \Leftrightarrow \exists \text{ a 1-form } \psi \text{ on } M \text{ s.t. } \Gamma'^{\mu}_{\nu\rho} = \Gamma^{\mu}_{\nu\rho} + \delta^{\mu}_{\nu}\psi_{\rho} + \delta^{\mu}_{\rho}\psi_{\nu}.$$

The equivalence between this algebraic definition of projective structure and the infinitesimal definition utilised by EPS follows from the facts that (i) two torsion-free affine connections agree on the same geodesics as unparameterised curves iff

$$\exists \text{ a 1-form } \psi \text{ on } M \text{ s.t. } \Gamma'^{\mu}_{\nu\rho} = \Gamma^{\mu}_{\nu\rho} + \delta^{\mu}_{\nu}\psi_{\rho} + \delta^{\mu}_{\rho}\psi_{\nu}$$

(see Eastwood (2008), proposition 2.1), and (ii) for any affine connection there can always be found a unique torsion-free affine connection that has the same geodesic structure. (In fact, as we will see later, EPS simply rule out torsionful affine structure by fiat!)

Finally, EPS' definition of metric structure is simply an informal statement of the claim that a (Lorentzian) metric g should be understood as the map $g : T_pM \times T_pM \to \mathbb{R}$ with suitable properties.

1.2 Construction of the differential topology

The first order of business for EPS is to construct differential-topological structure on M. As a starting point, EPS make the following assumption:[11]

> We accept in accordance with Axiom L_1 that there are 'figures' in M of the type shown in figure 4; i.e., a light signal L emitted from a particle P at p

[10] See e.g. Knebelman (1951) for details.

[11] Note that for better comparability with the original EPS paper, we use the same figure numbers as those used in that article.

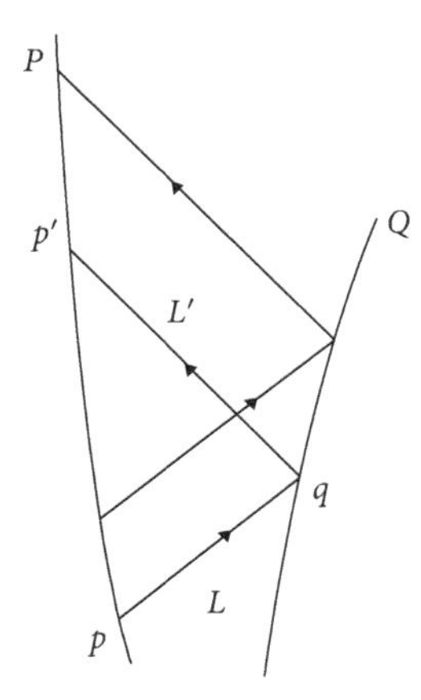

Figure 4 in Ehlers et al. (2012 [1972]) (own drawing)

towards another particle Q, where it is reflected at q and arrives back on P at P'. (Ehlers et al., 2012 [1972], p. 70)

Axiom L_1 will be discussed below. What is important for us here is the empirically-motivated posit that one can bounce a light ray from the world-line of one particle to that of another and back again. This is arguably the first (idealised) experimental statement in terms of $(M, \mathcal{P}, \mathcal{L})$ that is used to motivate structure on top of the triple $(M, \mathcal{P}, \mathcal{L})$, namely the notions of 'echo' and 'message':

Definition (Echo). *The map* $e_Q : P \to P, p \mapsto e_Q(p) =: p'$ *is called an* echo *on P from Q.*

Definition (Message). *The map* $m : P \to Q, p \mapsto m(p) =: q$ *is called a* message *from P to Q.*

In their next step, EPS take experimental observations relative to $(M, \mathcal{P}, \mathcal{L})$, *and* the notion of echos to justify the following assumption about the nature of particles $\mathcal{P}$ as well as echos between them:

Axiom (D_1). *Every particle is a smooth, one-dimensional manifold; for any pair P, Q of particles, any echo on P from Q is smooth and smoothly invertible.*

Axiom D_1 gives an ascription to particles as 'smooth, one dimensional manifolds', and to echos as being 'smooth and smoothly invertible'. We can thus read axiom D_1 as baking the empirically well-founded observations about the nature of particles and echos directly into (idealising) representational structures.

At this point, it is instructive to clarify more generally what is meant by a *constructivist*[12] approach, and how such an approach proceeds methodologically. A suitable definition of a constructivist system—perfectly compatible with the ambitions of EPS—is given in (Carnap, 1967, §2) (note that Carnap uses the word 'constructional' where we use 'constructivist'):

> To reduce [notion] *a* to [notions] *b*, *c* or to construct *a* out of *b*, *c* means to produce a general rule that indicates for each individual case how a statement about *a* must be transformed in order to yield a statement about *b*, *c*. This rule of translation we call a construction rule or constructional definition (it has the form of a definition; cf. §38). By a constructional system we mean a step-by-step ordering of objects in such a way that the objects of each level are constructed from those of the lower levels. Because of the transitivity of reducibility, all objects of the constructional system are thus indirectly constructed from objects of the first level. These *basic objects* form the *basis* of the system.

The basic objects on the EPS account can then be identified as given by a set of events, while novel structure is set up as follows ('construction takes place through definition' as Carnap (1967) would say):

1. Through definitions (such as that of 'echo' or 'message' just given).[13]
2. Through ascriptions of specific mathematical properties to and between structures already defined (the enrichment of the notion of echo by means of axiom D_1 is an example of this form)—accounting for (idealised) empirical observations. (The notion of 'particle' for instance gets ascribed the nature of a smooth, one-dimensional manifold. Although formally not completely correct, one can think of the notion of 'particle' as being thereby revised through the imposition of further structural desiderata; more strictly speaking, revision leads to a new notion of particle, say, 'particle*' in the sense of an explicit definition. Cf. (Leitgeb, 2011) on the notion of 'revision'.)

[12] Recall the distinction between 'constructivist' and 'constructive' already hinted at in the introduction to this chapter.

[13] This is partly related to Carnap's notion of 'definition in use' (Carnap, 1967, §§39-40), as well later notions of contextual definitions—see (Leitgeb, 2011) for further discussion.

Axioms of EPS are essentially instances of (2) but might also involve novel structure in terms of (1).

With all of this in mind, we proceed now to the next EPS axiom:

Axiom (D_2). *Any message from a particle P to another particle Q is smooth.*

Axiom D_2 gives an ascription to messages as 'smooth', baking the empirically found observation about messages directly into (idealising) representational structures.

Next, one introduces the notion of radar coordinates which will make it possible to be explicitly operationalist (rather than just constructive and constructivist):[14]

Definition (Radar coordinates). *Radar coordinates between two particles P and P', denoted as $x_{PP'}$, are charts on M for subset U such that a point $p \in U$ is coordinatised as (u, v, u', v') where u and v are respectively the emission and arrival times of light rays at P and u' and v' respectively the emission and arrival times of light rays at P'. A point on P specifically is coordinatised as (u, u, u', v') (since the emission and arrival times on P are identical). (See figure 5 in EPS, reproduced below.)*

(The operationalist underpinnings of radar coordinates are that one uses physical signal-and-mirror constructions to set up coordinates—this, for example, played a key role in Einstein's (1905) operational understanding of coordinates. As another example, think of how the sonar system in a submarine allows for operationalising the notion of distance through a technical procedure of emitting and receiving signals.)

Note that the definition on P is well-defined: u' and v' are determined uniquely by u; starting from u, one can determine u' and v' uniquely from u since there is—again anticipating axiom L_1—one unique line connecting the event at u on P and the event u' on P', and one unique line connecting the event at u on P and the event v' on P'.

The third axiom is as follows:

Axiom (D_3). *There exists a collection of triplets (U, P, P') where $U \subset M$ and $P, P' \in \mathcal{P}$ such that the system of maps $\{x_{PP'}|_U\}$ is a smooth atlas for M. Every other map $x_{QQ'}$ is smoothly related to the local coordinate systems of that atlas.*

[14] Arguably, axiom D_1 is already 'operationalist' (in a highly idealised sense) regarding the smoothness structures of the P.

Several technical remarks on axiom D_3 are in order:

1. Axiom D_3 is meant to turn M into a smooth manifold. From now on, M_p denotes the tangent (vector) space of M at p. However, a unique manifold is thus only determined if it is furthermore assumed that it satisfies the Hausdorff condition (see Lee (2022))—satisfaction of the Hausdorff condition is thus an implicit further assumption made by EPS![15]
2. According to EPS, axiom D_3 is a 'theorem' for Minkowski spacetime. The so-called theorem, however, follows trivially from the fact that Minkowski spacetime is (also) a manifold structure. The assumption of D_3 for the general case thus exhibits a sense in which some of the derived structure in Minkowski spacetime is still imposed—as an idealised empirical fact (more in the subsequent paragraph)—onto the more general spacetime structure by EPS—namely, M should still be (at least) a smooth manifold.
3. The second sentence in D_3 is a consistency claim: every other triple (U, Q, Q') not part of the atlas-generating collection is nevertheless found to be smoothly compatible with elements of that collection.
4. The coordinates of each $e \in M$ invoked in axiom D_3 are assigned as per figure 5.

As noted already, the EPS axioms capture experiential posits. Axiom D_3 is of course no exception: it can be understood as the concisely stated result from trying out radar coordinates, and thereby simply finding that these radar coordinate systems allow for tracking event space (and this so, in a partly compatible, partly complementary manner). It may strike one as problematic here (and, in fact, for the establishment of any of EPS' axioms) that the supposed empirical observations (say, about the compatibility and scope of radar coordinates) cannot be fully carried out in practice. (Most drastically, how would it ever be possible to span all of event space with a full atlas of radar coordinates?) So not only then do the axioms need to be thought of as idealising in the sense that they establish idealising representational structure (cf. our comment following axiom D_1 and D_2) but also their very own (demonstration of) validity is based on an assumption of generalisability and uniformity familiar from inductive reasoning. (More on this is §2.4.10.)

[15] Arguably, purely from what one is empirically entitled to say at this stage it seems in fact required to discard the Hausdorff condition—this, however, is not what EPS do. In order to avoid deviating too significantly from EPS' construction, we also simply assume satisfaction of the Hausdorff condition. To see what GR would look like without this condition, see Luc (2020); Luc and Placek (2020). We take up this point again in Chapter 3.

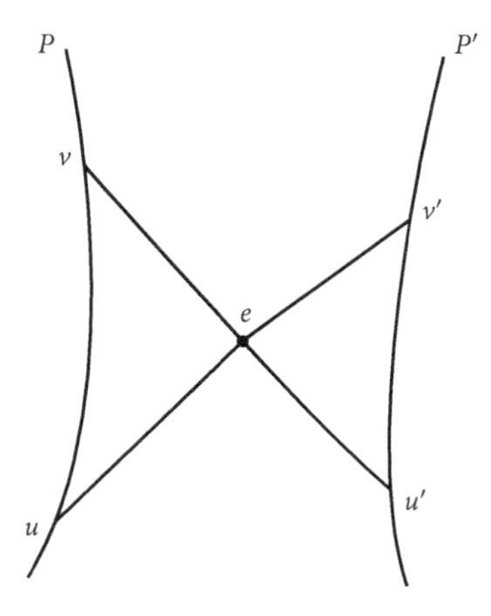

Figure 5 in Ehlers et al. (2012 [1972]) (own drawing)

Given the definitions and axioms presented thus far, EPS then make the following claim. Although they do not prove the claim explicitly, we provide a proof below.

Claim. *Every particle is a smooth curve in M.*

Before we continue to the proof, let us comment briefly on the nature of derived results such as claims/theorems/corollaries in EPS *vis-à-vis* that of axioms in a way that might seem self-evident but still strikes us as worth stressing. The claim on particles is helpful to this effect: the claim on particles does not ascribe a property to a particle but rather tells us about properties which follow from the nature of the particle as already established.

Proof. We divide this proof into numbered steps, for clarity:

1. *Claim:* Every particle P (a one-dimensional manifold) is also a one-dimensional smooth submanifold M' of M (with an atlas borrowed from M).

   Proof: A smooth submanifold is defined as follows:

   A subset $S \subset M$ is a k–dimensional *smooth submanifold* of M if and only if for every $p \in S$, there is a chart (ϕ, U, V) around p of M such that

   $$\phi(U \cap S) = V \cap (\mathbb{R}^k \times \{0\}) = \{x \in \phi(U) | x^{k+1} = \ldots. = x^n = 0\},$$

   where $U \subset M$, $V \subset \mathbb{R}^n$ and $n = \dim(M)$. (Wang 2020)

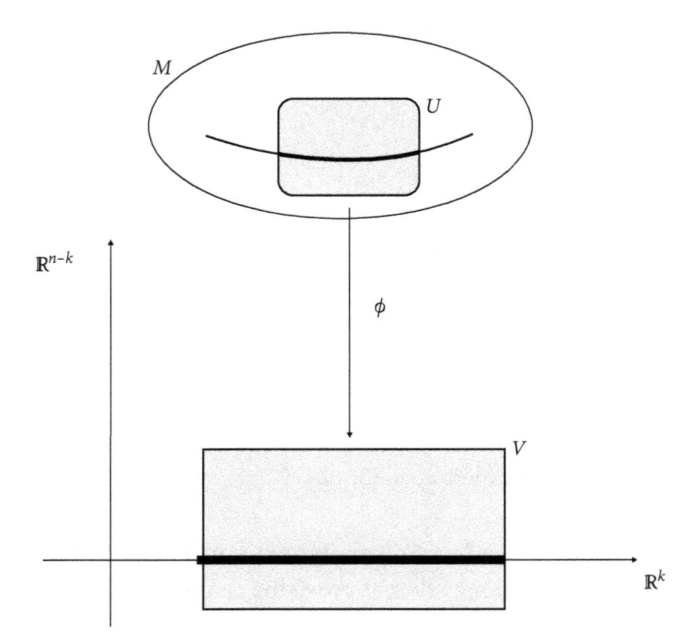

Figure Smooth submanifold definition (own drawing based on Wang (2020))

(This makes precise what is meant by the submanifold 'borrowing' an atlas.)

$x_{PP'}$ defines charts on M relative to two particles P and P' for some neighbourhood U; a point $p \in U$ is coordinatised as (u, v, u', v') where u and v are respectively the emission and arrival times at P, and u' and v' respectively the emission and arrival times at P'. A point on P specifically is coordinatised as (u, u, u', v') (since the emission and arrival times on P are identical). Note that u' and v' are uniquely determined by u: starting from P at u, one can uniquely determine the corresponding values of u' and v' through the unique line connecting the event at u on P and the events u', v' on P' respectively. We can thus switch to another coordinate system $\phi = (x^{(1)}, x^{(2)}, x^{(3)}, x^{(4)})$ which has all but one entry equal to zero on P ($x^{(2)} = x^{(3)} = x^{(4)} = 0$). But this means that for every $p \in P$ there is a chart (ϕ, U, V) around p of M—namely, the one just constructed—such that

$$\phi(U \cap P') = V \cap (\mathbb{R}^1 \times \{0\}^3) = \{x \in \phi(U) | x^{(2)} = \ldots. = x^{(4)} = 0\},$$

where $U \subset M$ and $V \subset \mathbb{R}^4$. Thus, every particle P (a one-dimensional manifold) is also a one-dimensional smooth submanifold M' of M.

2. Second, P is also a manifold in its own right; in fact, this is captured by axiom D_1 of EPS, but would follow now also from the fact that P is a submanifold of M. Then, the inclusion of $i : P \to M$ is smooth since $\phi \circ i \circ \psi^{-1} = j$ is smooth (where ϕ is a local chart on M—its existence follows from M being a manifold—, ψ is a local chart on P—its existence follows from P being a manifold in its own right—and $j : \mathbb{R}^1 \to \mathbb{R}^4 : x_1 \to (x_1, 0, 0, 0)$).[16]
3. Third, the map $g : \mathbb{R} \to P$ is smooth (axiom D_1).
4. The curve $i \circ g : \mathbb{R} \to M$ is smooth as i is smooth (step 2), and g is smooth (step 3). It is in this sense that we can say that a particle is a smooth curve in M. □

This claim established, there is one final EPS axiom of differential topology:

Axiom (D_4). *Every light ray is a smooth curve in M. If $m : p \to q$ is a message from P to Q, then the initial direction of L at p depends smoothly on p along P.*

There are some important questions which one might raise regarding the empirical status of axiom D_4. First: why assert that every light ray is a smooth curve in M, when the analogue for particles was derived from more fundamental axioms? The answer to this question is likely the following: in the case of particles, one can use the radar coordinate construction of figure 5, and the availability of two particles P and P', in order to derive this result. However, this radar coordinate construction is unavailable in the case of light rays, meaning that the assumption must be inserted 'by hand', by way of an axiom. (This is related to our inability to measure the one-way speed of light: just as we have no way of ascertaining the exact path traversed by a light ray in one direction, we likewise have no way of ascertaining whether this path is smooth. There is room, therefore, to argue that axiom D_4 is not in the strict spirit of a constructive axiomatic approach.[17] For more on foundational issues regarding measuring the one-way speed of light, see Salmon (1977).) One might also doubt that light rays (given the immense speed of light) can in any sense be 'observed' as contiguous sequences of events—which would after all

[16] Recall that a function between two manifolds is said to be smooth if and only if it is smooth in coordinates.

[17] Although against this, one might argue that since axiom D_4 involves no new definitional element, and is an empirical claim which is either true or false relative to the meaning of terms already introduced, it is thereby acceptable from the point of view of Reichenbachian constructive axiomatics. (Our thanks to Oliver Pooley for raising this point.)

be backing up the idealisation as continuous and, for EPS, even as smooth structures. But note that, for instance, the beams of laser pointers are easily made visible in the presence of, say, chalk dust; therefore, one might indeed accept light rays as visible just like particle trajectories.

The second question to be raised regarding axiom D_4—related to the first—is this: why has smoothness of light rays and particle trajectories been required so far, rather than simply n-times differentiability? Especially given that the scheme is to track observations, one might think that even a topological manifold simpliciter or even just a discrete lattice structure should be sufficient. It seems fair to say that the EPS scheme makes at this point a questionable idealisation not easily grounded in the empirical data alone. Lifting this assumption would lead to the possibility of the constructive axiomatization of, for example, low-regularity GR.[18]

1.3 Light propagation and conformal structure

With the differential-topological structure of M established, EPS now proceed to build up conformal structure from further axioms on the propagation of light. They begin with the following:

Axiom (L_1). *Any event e has a neighbourhood V such that each event p in V can be connected within V to a particle P by at most two light rays. Moreover, given such a neighbourhood and a particle P through e, there is another neighbourhood $U \subset V$ such that any event p in U can, in fact, be connected with P within V by precisely two light rays L_1, L_2, and these intersect P in two distinct events e_1, e_2 if $p \notin P$. If t is a coordinate function on $P \cap V$ with $t(e) = 0$, then $g(e) := -t(e_1)t(e_2)$ is a function of class C^2 on U.*

The geometrical constructions envisaged in axiom L_1 are presented in figure 6.

There are several remarks which must be made on axiom L_1. First, this axiom does not exclude any solution of general relativity, since such a neighbourhood exists locally for all such spacetimes. This is a result due to Perlick:[19]

[18] Cf. footnote 6.

[19] On the proof of the following proposition, Perlick writes:

> 'To prove this, we just have to recall that every point in a general-relativistic spacetime admits a convex normal neighborhood, i.e., a neighborhood V such that any two points in V can be connected by precisely one geodesic that stays within V. Having chosen such a V, it is easy to verify that every sufficiently small neighborhood U of p satisfies the desired property'. (Perlick, 2008, p. 134)

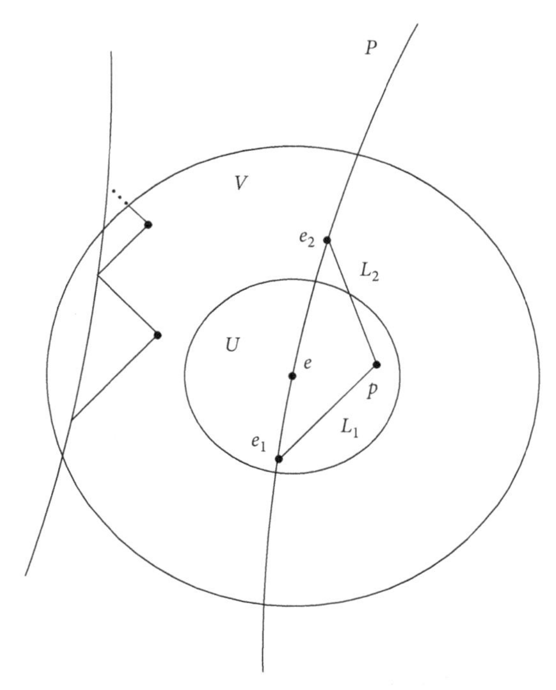

Figure 6 in Ehlers et al. (2012 [1972]) (own drawing)

Proposition (Perlick (2008), p. 4). *Let γ be a timelike worldine ('clock', in Perlick's terminology) in an arbitrary general relativistic spacetime and let $p = \gamma(t_0)$ be some point on γ. Then there are open subsets U and V of the spacetime with $p \in U \subset V$ such that every point q in $U \setminus image(\gamma)$ can be connected to the worldline of γ by precisely one future-pointing and precisely one past-pointing light ray that stays within V. In this case, U is called a* radar neighborhood *of p with respect to γ.*

Thus, EPS are not precluding the possibility of recovering the full solution space of general relativity by imposing axiom L_1. This, however, leads into our second remark on axiom L_1: as Pfister and King (2015) point out, this axiom cannot obtain globally in all solutions of general relativity:

> It should also be stressed here that, in strong gravitational fields, axiom L_1 is no longer valid globally: effects like the light deflection by big masses can lead to more than two light connections between p and P, and so-called horizons can have the effect that only one light connection exists, or none at all. Axioms L_1 and L_2 already contain a part of Einstein's equivalence principle,

so essential to general relativity: in gravitational fields, light has locally the same behavior as in special relativity. (Pfister and King, 2015, p. 56)

The third remark on axiom L_1 is this. The assumption that g, as defined in the axiom, is a function of class C^2 on U is (again) a strong one. Upon weakening this assumption, the EPS scheme can be liberalised to a constructive scheme for Finslerian spacetimes (of which Lorentzian spacetimes are a special case): see Lämmerzahl and Perlick (2018) and Pfeifer (2019). Consequently, it is implausible that Lorentzian spacetimes rather than Finslerian spacetimes are singled out from mere empirical considerations *à la* EPS.

These remarks being made, we now turn to the definition of directions and light-directions, and to axiom L_2:

Definition (Directions).

1. *The set of directions D_e at point e is the projective three-space associated with the tangent space M_e at e.*
2. *The set of light directions L_e is a subset of D_e.*

Analogous to how axioms $D_1, D_2, \ldots$ put flesh onto the bones of the notions of particles, message, and echo, L_2 ascribes properties to D_e, L_e and thus the tangent space as a whole:

Axiom (L_2). *The set L_e of light-directions at an (arbitrary) event e separates $D_e \setminus L_e$ into two connected components. In M_e the set of all non-vanishing vectors that are tangent to light rays consists of two connected components.*

Axiom L_2 establishes a purely point-based distinction between vectors, conceptually speaking a point-based predecessor to a distinction between space and time[20] as well as future and past. (It is perhaps more straightforward to understand the distinction brought in by L_e into D_e as one into four connected components, two components for what will become 'spacelike' sectors, and two components for what will become the future and the past timelike sectors respectively.) Note also that at least a (locally imposed) conformal connection would be needed as a surplus structure to 'continuously' link the thus-established distinction at the tangent space of one point to that at the tangent space of a neighbouring point.

[20] Cf. (Le Bihan and Linnemann, 2019).

All axioms are now in place to allow for the construction of conformal structure. Schematically, the construction has the following intermediate steps:

1. Explore the mathematical properties of g. In particular, construct a rank-two tensor $g_{\mu\nu}$ from g.
2. Use observational data to select a preferred non-degenerate quadric associated with $g_{\mu\nu}$.[21] This non-degenerate quadric will express the signature on M as well as provide a threefold distinction of the elements of D_e (extending the distinction between vectors introduced with axiom L_2).
3. Show that this quadric can be expressed in terms of a tensor density.

On (1), EPS begin with a lemma, stated and proved below:

Lemma (Properties of g). *The following statements hold about g:*

(a) $g(p) = 0$ *(for $p \in U$) if and only if p lies on a light ray through e.*
(b) $g_{,\mu}(e) := \partial_\mu g(e) = 0$.
(c) $g_{\mu\nu}(e) := g_{,\mu\nu}(e) := \partial_\mu\partial_\nu g(e)$ *defines a tensor at e.*
(d) The tangent vector T^μ of any light ray L through the point e satisfies $g_{\mu\nu}T^\mu T^\nu = 0$.
(e) $g_{\mu\nu} \neq 0$ *on particles.*

Proof. ad (a): Left-to-right: Suppose $g(p) = -t(e_1)\,t(e_2) = 0$. Thus, either $t(e_1) = 0$ or $t(e_2) = 0$. $t : V \cap P \to \mathbb{R}$ is a coordinate function with $t(e) = 0$; since it is a coordinate function, it has to be strictly monotonic and so cannot be zero at any other event than e. Hence, either $e_1 = e$ or $e_2 = e$. So p lies on a light ray through e. Right-to-left: $g(p) = -t(e_1)\,t(e_2)$. If p lies on a light ray through e, then pick $e_1 = e$ (at $t(e) = 0$). Thus, $g(p) = -t(e)\,t(e_2) = 0$.

ad (b): Suppose for contradiction that $g_{,\mu}(e) \neq 0$.[22] Consider the (local) hypersurface picked out by the condition $g(p) = 0$ on the points p in the

[21] Generally, a quadric is a D-dimensional hypersurface in a $(D + 1)$-dimensional space that is determined by the zero set of an irreducible second degree polynomial in $D + 1$ variables (i.e. a corresponding quadratic form). Notably, in the following, quadrics are to be understood as infinitesimal quadrics in direct analogy with the infinitesimal cone structure associated with conformal structure (introduced earlier). So, just as the infinitesimal cone structure on the manifold can be expressed algebraically on the tangent bundle, so too can the infinitesimal quadric structure be so expressed.

[22] In deploying here a proof by contradiction, we (like EPS) explicitly do not commit to a constructive approach to *mathematics*, but only to the axiomatization of our physical theories, in the sense already articulated above. The relation between constructive axiomatics and constructive mathematics will be discussed further in Chapter 2.

neighbourhood U of e. This is in fact a smooth hypersurface around e since the normal $g_{,\mu}(e) \neq 0$ is non-zero by assumption (in coordinates).[23] Understood via the inverse of the exponential map (a smooth function), one can think of the surface as a smooth surface, cutting tangent space into two. This would, however, contradict the second part of axiom L_2, which states that the set of all non-vanishing vectors that are tangent to light rays should consist of four connected components:[24] this cannot be the case if the hypersurface in question is smooth as the tangent space at e is then the standard tangent space of a smooth (sub)manifold, which is in particular connected. Thus, it must be the case that $g_{,\mu}(e) = 0$.

ad (c): One can demonstrate that $g_{\mu\nu}(e)$ has tensorial transformation properties—and therefore defines a tensor at e—as follows:

$$\begin{aligned} g_{\mu\nu} &:= \frac{\partial^2 g}{\partial x^\mu \partial x^\nu} = \frac{\partial}{\partial x^\mu}\left(\frac{\partial}{\partial x^\nu} g\right) \\ &\mapsto \frac{\partial x^{\mu'}}{\partial x^\mu}\frac{\partial}{\partial x^{\mu'}}\left(\frac{\partial x^{\nu'}}{\partial x^\nu}\frac{\partial}{\partial x^{\nu'}} g\right) \\ &= \frac{\partial x^{\mu'}}{\partial x^\mu}\frac{\partial^2 x^{\nu'}}{\partial x^{\mu'}\partial x^\nu}\frac{\partial}{\partial x^\nu} g + \frac{\partial x^{\mu'}}{\partial x^\mu}\frac{\partial x^{\nu'}}{\partial x^\nu}\frac{\partial^2}{\partial x^{\mu'}\partial x^{\nu'}} g \\ &= \frac{\partial x^{\mu'}}{\partial x^\mu}\frac{\partial x^{\nu'}}{\partial x^\nu} g_{\mu'\nu'}. \end{aligned}$$

The first term vanishes in the above since $\partial_\mu g(e) = 0$. So $g_{\mu\nu}$ has the requisite tensorial transformation properties, and therefore defines a tensor at e.

ad (d): Differentiating twice with respect to s from $g(x^\mu(s)) = 0$, we have:

$$\begin{aligned} \frac{\partial g}{\partial x^\mu}\frac{dx^\mu}{ds} &= 0 \\ \Rightarrow \frac{d}{ds}\left[\frac{\partial g}{\partial x^\mu}\frac{dx^\mu}{ds}\right] &= 0 \\ \Rightarrow \frac{\partial^2 g}{\partial x^\mu \partial x^\nu}\frac{dx^\mu}{ds}\frac{dx^\nu}{ds} + \frac{\partial g}{\partial x^\mu}\frac{d^2 x^\mu}{ds^2} &= 0 \\ \Rightarrow g_{\mu\nu} T^\mu T^\nu &= 0. \end{aligned}$$

[23] The normal to a hypersurface being non-zero is sufficient to render it smooth. This follows from the regular set theorem (see Tu (2011), §9.4); the remark under Lemma 9.10 demonstrates nicely that a non-zero normal is not necessary for smoothness. Heuristically, the sufficiency of a non-zero normal for hypersurface smoothness should seem natural from the fact that the implicit function theorem can be used to establish the smoothness of a level set (such as that one defining a hypersurface).

[24] See our elaborations below axiom L_2 for the sense in which there are *four* components here.

where here $T^\mu := \frac{dx^\mu}{ds}$ and again we have used that $\partial_\mu g(e) = 0$.

ad (e): If $x^\mu(t)$ is the parameter representation of P in any permissible coordinate system, then $g\{x^\mu(t)\} = -t^2$; hence, $g_{\mu\nu}K^\mu K^\nu = \partial_t^2(-t^2) = -2$, K^μ being the tangent vector of P at e with respect to t (so $K := \partial/\partial t$). Hence, $g_{\mu\nu} \neq 0$ on particles. □

This lemma established, we turn now to step (2): associating a preferred quadric with $g_{\mu\nu}$. We first prove another lemma:

Lemma ($L_e \subseteq Q$). *The set of light rays at event e, L_e, is contained in the 3-dimensional quadric Q defined in D_e by the quadratic form $g_{\mu\nu}T^\mu T^\nu = 0$.*

Proof. This result follows from the fact that light rays on e with tangent T^μ fulfil $g_{\mu\nu}T^\mu T^\nu = 0$ (see lemma (properties of g, part (d))). □

We normalise $g_{\mu\nu}$ (new notation: $\mathscr{g}_{\mu\nu}$) such that $\det(\mathscr{g}_{\mu\nu}) = -1$; this leaves the definition of the quadric form Q unaffected ($\mathscr{g}_{\mu\nu}$ induces the same form as $g_{\mu\nu}$). It can then be shown (as the announced step (3)) that:

Claim. $\mathscr{g}_{\mu\nu}$ *is a tensor density.*

Proof. We have already seen that $g_{\mu\nu}$ defines a tensor. $\mathscr{g}_{\mu\nu}$ is a normalised version of g_{ab}, transforming as a tensor up to a multiple of the metric determinant. It is therefore a tensor density, in the sense of (Anderson, 1967, §1.7) (cf. (Bergmann, 1942, ch. 16)). □

Using $\mathscr{g}_{\mu\nu}$, we can then distinguish between different classes of vectors as follows:

Definition (Vector distinctions). *A vector T^μ is called ($\mathscr{g}$-)null/timelike/spacelike if and only if $\mathscr{g}_{\mu\nu}T^\mu T^\nu$ is, respectively, equal to, greater than, or less than zero.*

One can then prove the following claim:[25]

Claim (Properties of Q and $L_e \subseteq Q$). *The (general) quadratic form associated with the quadric Q is (a) non-degenerate (there is no $X^\mu \neq 0$ such that, for all*

[25] Note that in EPS the definition of the tensor density follows after mentioning this claim; we, however, wish the vector distinction—induced from the quadric Q/the tensor density $\mathscr{g}$—to be already available at this stage in order to show the non-degeneracy of the quadric Q.

Y^μ, $g_{\mu\nu}X^\mu Y^\nu = 0$), and (b) normal hyperbolic (i.e., the quadric is of signature $(+,+,+,-)$). Furthermore, (c) the set of light-directions is equal to the quadric, i.e. $L_e = Q$.

Proof. *ad (a)*: Suppose for contradiction that the quadric form is degenerate. Given that $g_{\mu\nu}$ is a symmetric bilinear form by construction, this is equivalent to $g_{\mu\nu}$ being singular. In particular, in some coordinate system $\{e_i\}$ ($i = 1, \ldots, n$, where n is the dimension of the manifold), $g_{\mu\nu}$ can be diagonalised with at least one zero on the diagonal. This, however, means that relative to this coordinate system, we can write the quadric equation without at least one of the coordinates. Thereby, the level set is revealed to be of codimension higher than 1, and in particular not a three-dimensional hypersurface (if one is working, as EPS are, in the case $n = 4$) that can induce a distinction in directional space for L_e. But exactly this is required by axiom L_2 and thus also required for Q to capture consistently the structure of light propagation.

ad (b): Assuming non-degeneracy, (Synge, 1965, pp. 17–18) demonstrates that only the $(+,+,+,-)$ quadric signature can separate future and past into two connected components.[26] 'Normal hyperbolicity' just means that the signature of the quadric is $(+,+,+,-)$.[27]

ad (c): No proper subset of any quadric has the required topological properties of separating future and past into two connected components, so L_e must coincide with the entire quadric, yielding $L_e = Q$. To see this, assume that L_e is a proper subset of Q while still being a quadric. But removing even a single point from Q means that L_e will not allow for splitting the space-like from the time-like directions any more—in conflict with the requirement on L_e from the first part of axiom L_2. □

After this setup, EPS can take the resulting $\mathscr{g}_{\mu\nu}$ to define a conformal structure (denoted by $\mathcal{C}$ from now on): $\mathscr{g}_{\mu\nu}$ stands for an equivalence class of $g_{\mu\nu}$ (see the construction of $\mathscr{g}_{\mu\nu}$ above), which itself defines a conformal structure $\mathcal{C}$ in the algebraic sense introduced in section 1.1.

One defines (note that this is essentially the same definition of ($\mathscr{g}$-)null/timelike/spacelike as given above; we only deploy the following definition so that our notation aligns with that of EPS):

Definition (Vector distinctions (2)). *A vector T^μ is called ($\mathcal{C}$-)null/timelike/spacelike if and only if $\mathscr{g}_{\mu\nu}T^\mu T^\nu$ is, respectively, equal to, greater than, or less than zero.*

[26] Recall that the definition of the signature of a quadric just is the structure of + and – signs in an algebraic representation of said quadric, i.e. its quadratic form.

[27] Note that order and denotation of + as opposed to – is subject to convention; other conventional choices would fix the same quadric.

Explicitly, $\mathcal{C}$ is set up via an operational procedure for determining the components of $\mathcal{g}_{\mu\nu}$ relative to some coordinate system which proceeds as follows:

Claim. *There exists a basic operational procedure to determine values of the conformal structure* $\mathcal{C}$ *associated to* $\mathcal{g}_{\mu\nu}$ *relative to a radar coordinate system for a point e.*

Proof. The components of $\mathcal{g}_{\mu\nu}$ relative to a coordinate system can be ascertained via the following steps:

1. Note first that nine equations in nine unknowns would be sufficient to fix the nine components of $\mathcal{g}_{\mu\nu}$ (given that $\mathcal{g}_{\mu\nu}$ has one less degree of freedom than a metric $g_{\mu\nu}$).
2. Create a light ray for each direction (hereby, use the notion of linear independence provided by a coordinate basis in order to identify distinct directions). There are three directions.
3. Use these three linearly independent light rays to construct nine linearly independent squares which fix $\mathcal{g}_{\mu\nu}$: the light ray in direction i will be made to hit a point p_i sufficiently close to the emission point so that it can still be coordinatised in radar coordinates. Once the radar coordinates for p_i are known, the corresponding directional vector can be expressed in the tangent space coordinates dual to the radar coordinates (i.e. $\partial/\partial T$, $\partial/\partial X$, $\partial/\partial Y$, $\partial/\partial Z$ with T, X, Y, Z being the radar coordinates).[28] □

Clearly, light rays are $\mathcal{C}$-null curves (this follows directly from part (d) of lemma (properties of g)). We want to show now, though, that light rays are $\mathcal{C}$-null *geodesics*. The other major goal is to establish that (local) light ray cones are captured by the $\mathcal{C}$-null cone structure.[29] Towards these goals, EPS first need to establish some intermediate results:

Lemma. $v_e := \{q \in M : g(q) = 0, q$ *is light ray connected to e*$\}$ *is a smooth hypersurface, except at e itself.*

[28] Practically, it needs to be ensured that the light ray gets re-emitted at p_i in the direction of the observer. This could for instance be guaranteed by simply re-emitting into all directions, together with a label on the signals that one derives from signals in direction i (which must then in turn have, of course, been equipped with a label of its direction as well). More concretely, the function of the label could be realised by the frequency of the light ray chosen, and the function of the re-emitter by an antenna tuned only to that frequency.

[29] Remember that conformal structure can be both be characterised algebraically and geometrically, as explained in section 1.1.

Proof. First show that there exists a neighbourhood W of e such that if $e \neq q \in W$ and $g(q) = 0$, then $g_{,\mu}(q) \neq 0$. For this, note that

$$g_{,\mu}(q) - \underbrace{g_{,\mu}(e)}_{=0 \text{ as shown in part (b) of lemma (Properties of } g)} = \int_e^q g_{\mu\nu}\{x^\lambda(u)\} \underbrace{T^\mu(u)}_{:=\frac{dx^\mu}{du}} .$$

In particular, at e, $g_{\mu\nu}T^\nu \neq 0$; the latter function is continuous in u. Thus, for sufficiently small departure from e (in u alternatively speaking), $g_{\mu\nu}T^\nu \neq 0$ as well due to continuity (use here that the set of initial directions is compact and thus bounded). Thus,

$$g_{,\mu}(q) = \int_e^q \underbrace{g_{\mu\nu}T^\nu}_{:=f(u)} du = \int_e^q \underbrace{f(u)}_{\neq 0} du = \underbrace{f(c)}_{\neq 0} \underbrace{[u(q) - u(e)]}_{\neq = 0 \text{ for } q \neq e} \neq 0.$$

where we used the mean value theorem.

Second, read $g(q) = g(x^\lambda(q)) = g(x^\lambda(q), q)$, and $g_{,\mu}(q) = g_{,\mu}(x^\lambda(q), q)$ (i.e., through their definition in coordinates) where $g, g_{,\mu} : W'xW \subset (\mathbb{R}^4 \times M) \to \mathbb{R}$. Together with the fact that $g(q) = 0$ and the result that $g_{,\mu}(q) \neq 0$, the implicit function theorem then tells us that there exists a unique assignment $U \subset M \to \mathbb{R}^4, q \mapsto x^\lambda(q)$ such that $g(x^\lambda(q), q) = 0$. But this in turn means that ν_e is a smooth manifold, i.e. a smooth hypersurface near e because, for every q in ν_e, there is a neighbourhood U now so that the mapping $U \to \mathbb{R}^4$ is smooth (this thus provides for an atlas on ν_e). Finally, recall that it has already been shown in course of the proof of part (b) of lemma (Properties of g) that there cannot be a smooth hypersurface at e. □

Claim. *(Null hypersurfaces) The integral curves associated to the null vector field K that generates a null hypersurface S are $\mathfrak{C}$-null geodesics.*

Proof. We need to show that for such a vector field K, the geodesic equation (up to reparameterisation) $\nabla_K K = \lambda K$ obtains. Importantly, this statement follows from the fact that at each $p \in S$, we have $\nabla_K K \perp T_p S$ (call this condition $(*)$), that is $\mathscr{g}(\nabla_K K, X) = 0$ for all $X \in T_p S$. (After all, if $\nabla_K K$ is orthogonal to the hypersurface S, it can only be parallel to its defining orthogonal vector K.) We will now show $(*)$ and thus establish the claim.

First, extend the arbitrary tangent vector $X \in T_p S$ via a flow induced by K such that X is invariant under that flow, i.e. $[K, X] := \nabla_K X - \nabla_X K = 0$. As X remains tangent to $T_p S$ along the flow generated by K, we have $\mathscr{g}(X, K) = 0$.[30]

[30] We follow up to this point the proof of Galloway (2004), proposition 3.1.

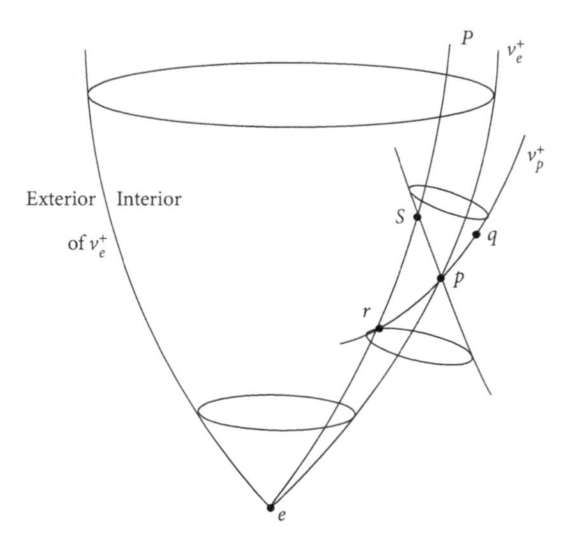

Figure 7 in Ehlers et al. (2012 [1972]) (own drawing)

One now differentiates with respect to K:

$$0 = K(\mathscr{g}(K, X)) = (\nabla_K \mathscr{g})(K, X) + \mathscr{g}(\nabla_K K, X) + \mathscr{g}(K, \nabla_K X).$$

Now, a conformal connection can be induced from the Levi-Civita connection associated with an arbitrary element of a conformal equivalence class of metrics (this follows in particular in the current context from the algebraic definition of conformal structure of $\mathcal{C}$); for such a conformal connection, one has $\nabla_\mu \mathscr{g}_{\nu\lambda} = 0$ (Curry and Gover, 2018, p. 17). Thus, the above simplifies to $0 = \mathscr{g}(\nabla_K K, X) + \mathscr{g}(K, \nabla_K X)$, and hence

$$\mathscr{g}(\nabla_K K, X) = -\mathscr{g}(K, \nabla_K X) = -\mathscr{g}(K, \nabla_X K) = -\frac{1}{2}X(\mathscr{g}(K, K)) = 0,$$

where we used $[K, X] = 0$ (as established above). □

Proposition (Characterisation of conformal structure $\mathcal{C}$).

1. *Light rays are $\mathcal{C}$-null geodesics.*
2. *ν_e is a $\mathcal{C}$-null cone at $e \in M$.*
3. *The map $e \mapsto \nu_e$ establishes a conformal structure on M.*

Proof. Choose a neighbourhood around a point e such that (1) this neighbourhood is a V-type neighbourhood (see axiom L_1), and (2) it admits a timelike vector field, i.e. the neighbourhood is a time-orientable neighbourhood. (The existence of individual neighbourhoods of this kind is guaranteed from the previous claim and L_1; the chosen neighbourhood is just their joint intersection.)

Let p be an event on the future light cone emanating from e (itself denoted by v_e^+), within the V neighbourhood. Without loss of generality, we can furthermore restrict V to a neighbourhood as described in axiom C, i.e. for each event p in that neighbourhood, there is a particle through e that also goes through p if and only if p lies in the interior of v_e.

The tangent space on the light cone v_e^+ at point $p \in V$ cannot be spacelike: if so, the cone could not completely be filled with particles as, however, required by axiom C (for recall from lemma (Properties of $\mathcal{P}$-geodesics, 1) that particles are nowhere spacelike).

Furthermore, the tangent space at p on v_e^+ cannot be timelike: assume that the tangent space is timelike. Consequently, the light cone v_p^+ will intersect—and not align with—v_e^+ at p. Note then that one can find a light ray (r, p, q) that goes from some point r in the interior of v_e^+ via p on v_e^+ to some point q in the exterior of v_e^+ (see figure 7). By choosing r close enough to p, and using that according to axiom C the interior of v_e^+ is filled up locally with particles passing through e,[31] one can find a particle P from e through r that is connectible to p through at least three (rather than exactly two) light rays (in clash with the initial V neighbourhood assumption): by the two light rays arising from the intersection of the backward cone v_p^- with v_e^+, and (at least) a light ray going from p to s (the latter is guaranteed from taking a V neighbourhood within the assumed V neighbourhood, that, however, excludes e).

This means then that the tangent space $T_p v_e^+$ for any p in the specified neighbourhood of e is null. Thus, v_e^+ is a smooth *null* hypersurface near e. As the null vector field associated to the light rays generates the hypersurface,[32] they must then be, in virtue of the proven claim (Null hypersurfaces), null geodesics. In particular, the local light cone v_e is then indeed identical with the $\mathcal{C}$-null cone at e. □

In summary: EPS' methodology here has been to first construct a conformal structure $\mathcal{C}$ in turn inducing a tripartite classification of vectors at each

[31] In fact, EPS reference here axiom P_2 but it is clear that axiom C must have been meant.

[32] Note that integral curves to a null vector field need not generate a hypersurface that is null; this follows from the claim (Null hypersurfaces) according to which integral curves of a null vector field linked to a null hypersurface are null geodesics, and the simple fact that not all null curves that are integral curves of a null vector field are null geodesics! However, we have proven that the generated hypersurface is indeed a null hypersurface.

point as timelike/spacelike/null, then to show that the cone of light rays ν_e is a $\mathcal{C}$-null cone, and thereby to show that the map $e \mapsto \nu_e$ establishes a conformal structure on M. Now, one might think that since we have already seen that EPS define a conformal structure 'intrinsically' (rather than 'algebraically') as a field of null cones over M, it is not entirely obvious that the methodology followed by EPS constitutes the most direct means of establishing a conformal structure on M via the causal-inertial constructivist approach. Note, though, that one is after showing with the above constructions that light ray cones are exactly these *null* cones (assuming that one cannot simply define the cones given by each ν_e to be null.)

1.4 Free fall and projective structure

Having constructed a conformal structure $\mathcal{C}$ from axioms regarding the behaviour of light rays, EPS turn next to the construction of projective structure. They begin with two axioms:

Axiom (P_1). *Given an event e and a $\mathcal{C}$-timelike direction D at e, there exists one and only one particle P passing through e with direction D.*

Axiom (P_2). *For each event $e \in M$, there exists a coordinate system $(\overline{x}^\mu)$, defined in a neighbourhood of e and permitted by the differential structure introduced in axiom D_3, such that any particle P through e has a parameter representation $\overline{x}^\mu(\overline{u})$ with*

$$\left.\frac{d^2\overline{x}^\mu}{d\overline{u}^2}\right|_e = 0;$$

such a coordinate system is said to be projective at e.

Regarding axiom P_2, Sklar (1977) raises the concern that this presupposes an understanding of which particles are forced versus force-free—since the force-free particles will satisfy the equation stipulated in the axiom, while forced particles will not—and that this understanding thereby transcends the empiricist motivations of EPS. There are two central lines of response here. The first is to accept that this a conventional, rather than empirically motivated, input in the EPS scheme. The second is to argue that we are not, after all, free to stipulate which particles are forced versus force-free: for example, Coleman and Korte (1980) argue that the distinction can be *derived* merely from differential-topological inputs of the kind already discussed above. The issues here are delicate—for example, Pitts (2016) responds that the Coleman-Korté line of reasoning does not succeed. Even though we ultimately side with Coleman-Korté's apporach, suffice it here simply to register that this is a point of controversy: these matters are discussed in greater detail in Chapter 2.

Note also that if the set of allowed coordinate charts does not exclude holonomic coordinates, then torsion-freeness does not follow. EPS appear to assume implicitly that coordinate systems *are* holonomic (this will become particularly evident in the following corollary), thereby excluding torsionful spacetimes by fiat. (For further discussion on anholonomic coordinates, see (Iliev, 2006) and (Hartley, 1995), as well as the philosophical assessment by Knox (2013).)

Corollary. *In an arbitrary coordinate system, one has*

$$\ddot{x}^{\mu} + \Pi^{\mu}_{\nu\lambda}\dot{x}^{\nu}\dot{x}^{\lambda} = \lambda\dot{x}^{\mu}, \tag{1.1}$$

where λ depends on the choice of u and where one can restrict $\Pi^{\mu}_{\nu\lambda}$ to $\Pi^{\mu}_{[\nu\lambda]} = 0$ and $\Pi^{\mu}_{\nu\mu} = 0$.

Note here that restriction to symmetric connection coefficients rules out torsionful spacetimes, as discussed above. Furthermore, note that imposing $\Pi^{\mu}_{\nu\mu} = 0$ is always justified for a projective connection, since for any projective connection for which this does not hold, one can always find an equivalent projective connection for which the condition does hold (i.e. the equivalent connection defines the same paths)—see (Crampin and Saunders, 2007, p. 697).

Definition (Projective structure (EPS)). *The structure imposed using $\Pi^{\mu}_{\nu\lambda}$ is known as a* projective structure *$\mathcal{P}$: a family of curves, called geodesics, whose members behave in the second-order infinitesimal of each neighbourhood of M like the straight lines of an ordinary projective four-space.*

Definition (Geodesic). *Any curve that satisfies (1.1) is said to be ($\mathcal{P}$)-*geodesic.

Proposition (Uniqueness of projective structure). *The coefficients $\Pi^{\mu}_{\nu\lambda}$ are determined uniquely at e for any particular coordinate system $\{x^{\mu}\}$ around e.*

Proof. Suppose there is another projective structure $\overline{\Pi}^{\mu}_{\nu\lambda}$ determined at e in this coordinate system by an analogous equation to (1.1), that is

$$\ddot{x}^{\mu} + \overline{\Pi}^{\mu}_{\nu\lambda}\dot{x}^{\nu}\dot{x}^{\lambda} = \lambda\dot{x}^{\mu}.$$

Denote the difference between $\overline{\Pi}^{\mu}_{\nu\lambda}$ and $\Pi^{\mu}_{\nu\lambda}$ as $\tilde{\Delta}^{\mu}_{\nu\lambda}$. Note that $\tilde{\Delta}^{\mu}_{\nu\lambda}$ is symmetric in its lower two indices, since $\overline{\Pi}^{\mu}_{\nu\lambda}$ and $\Pi^{\mu}_{\nu\lambda}$ are both also symmetric in their lower two indices.

Now, axiom P_1 and equation (1.1) imply $T^{[\lambda}\tilde{\Delta}^{\mu]}{}_{\nu\rho}T^{\nu}T^{\rho} = 0$ where $T^{\lambda} := \dot{x}^{\lambda}$, i.e. is the tangent to the curve. To derive $\tilde{\Delta}^{\lambda}_{\mu\rho} = 0$, proceed as follows.

For all T^{μ}, it holds that

$$
\begin{aligned}
T^{[\lambda}\tilde{\Delta}^{\mu]}{}_{\nu\rho}T^{\nu}T^{\rho} &= 0\\
\Rightarrow \left(T^{\lambda}\tilde{\Delta}^{\mu}_{(\nu\rho)} - T^{\mu}\tilde{\Delta}^{\lambda}_{(\nu\rho)}\right)T^{\nu}T^{\rho} &= 0\\
\Rightarrow T^{\lambda}\tilde{\Delta}^{\mu}_{(\nu\rho)} - T^{\mu}\tilde{\Delta}^{\lambda}_{(\nu\rho)} &= \Xi^{\lambda\mu}_{(\nu\rho)} \quad \text{and} \quad \Xi^{\lambda\mu}_{(\nu\rho)}T^{\nu}T^{\rho} = 0\\
\Rightarrow T^{\lambda}\tilde{\Delta}^{\mu}_{(\nu\rho)} - T^{\mu}\tilde{\Delta}^{\lambda}_{(\nu\rho)} &= 0\\
\Rightarrow \tilde{\Delta}^{\mu}_{(\nu\rho)} &= 0
\end{aligned}
$$

The second to last step follows from the fact that $\Xi^{\lambda\mu}_{(\nu\rho)}$ would have to be linear in T^{μ} or T^{λ}, thus $\Xi^{\lambda\mu}_{(\nu\rho)} = \alpha T^{\lambda}\Xi'^{\mu}_{(\nu\rho)} + \beta T^{\mu}\Xi''^{\lambda}_{(\nu\rho)}$ for some factors α, β. Now, for any T^{μ}, the second equation in the third line above becomes $\alpha\Xi'^{\mu}_{(\nu\rho)}T^{\lambda}T^{\nu}T^{\rho} + \beta\Xi''^{\lambda}_{(\nu\rho)}T^{\mu}T^{\nu}T^{\rho} = 0$, with $\Xi'^{\mu}_{(\nu\rho)}, \Xi''^{\mu}_{(\nu\rho)}$ independent of T^{μ}. From that $\Xi'^{\mu}_{(\nu\rho)}, \Xi''^{\mu}_{(\nu\rho)}$ are independent of T^{μ}, we can infer that $\Xi'^{\mu}_{(\nu\rho)}, \Xi''^{\mu}_{(\nu\rho)}$ must be zero, in which case the penultimate line follows. The last step follows from that $\Delta^{\mu}_{(\nu\rho)}$ is independent of T^{μ}.

Thus, the antisymmetric and symmetric parts both vanish, and we have $\tilde{\Delta}^{\mu}_{\nu\lambda} = 0$. □

Claim. *The coefficients $\Pi^{\mu}_{\nu\lambda}$ can be measured in radar coordinates.*

Proof. First, manipulate (1.1) by contracting it with another tangent vector while antisymmetrizing it:

$$
\begin{aligned}
\ddot{x}^{\mu} + \Pi^{\mu}_{\nu\lambda}\dot{x}^{\nu}\dot{x}^{\lambda} &= \lambda\dot{x}^{\mu}\\
\Rightarrow \dot{x}^{[\mu}(\ddot{x}^{\sigma]} + \Pi^{\sigma]}_{\nu\lambda}\dot{x}^{\nu}\dot{x}^{\lambda}) &= \lambda\dot{x}^{[\mu}\dot{x}^{\sigma]}\\
\Rightarrow \dot{x}^{[\mu}(\ddot{x}^{\sigma]} + \Pi^{\sigma]}_{\nu\lambda}\dot{x}^{\nu}\dot{x}^{\lambda}) &= 0. \qquad (1.2)
\end{aligned}
$$

Note now that all $x^{\mu}, \dot{x}^{\mu}$, etc. can in principle be constructed using radar coordinates, while the parameter λ—which is not directly determinable empirically—has been removed. □

Thus, EPS have (i) derived a projective structure $\mathcal{P}$ from the salient empirically motivated axioms, (ii) demonstrated the uniqueness of this structure, and (iii) demonstrated how this structure can be measured operationally using radar coordinates.

1.5 Compatibility of free fall and light propagation: the construction of Weyl structure

Having constructed a conformal structure $\mathcal{C}$ from the trajectories of light rays, and a projective structure $\mathcal{P}$ from the trajectories of freely falling particles, EPS now introduce a 'compatibility' axiom in order to build a Weyl structure from these structures. This compatibility axiom is stated as follows:

Axiom (*C*). *Each event e has a neighbourhood U such that an event $p \in U$, $p \neq e$ lies on a particle P through e if and only if p is contained in the interior of the light cone ν_e of e.*

In other words: For all events e, there exists a neighbourhood U such that for all events $p \in U$ the following equivalence holds: there exists a particle P with $p, e \in P$ iff p lies in the interior of ν_e.

We can think of axiom C as the observational statement that particles always move within the light-cone structure. Axiom C is thereby reminiscent of the relativity postulate in special relativity, which, understood as a 'principle', also has the status of an empirically backed result. (We discuss further the connections between constructive axiomatics and principle theories of physics in Chapter 2.)

From axiom C, EPS prove the following lemma:

Lemma (Properties of $\mathcal{P}$-geodesics)

1. *Every particle is a $\mathcal{P}$-geodesic which is nowhere $\mathcal{C}$-spacelike.*
2. *Every $\mathcal{P}$-geodesic that is timelike at some event can nowhere be spacelike.*

Proof. ad (1): That every particle is a $\mathcal{P}$-geodesic follows from axiom P_2. Assume for contradiction that the $\mathcal{P}$-geodesic describing a particle P is spacelike at e. Tangent vectors, via the exponential map, uniquely define up to a certain parameter value r curves that do not intersect with one another; in particular, the curve γ associated via the (inverse) exponential map to the spacelike tangent at e to P (with $\{\gamma(t) : t \in [0, r)\} \subset P$) does not intersect with any curve η associated to a null vector at e as long as $r < r'$. Now, choose an s such that $0 < s < r$. It then follows that the point $p_s := \gamma(s)$ will be on P but not in the interior of ν_e (this would require having intersected with η).

But then there exists an event (namely e) such that for all its neighbourhoods U there exists $p \in U$ (namely p_s with sufficiently small $s > 0$ so that p_s is still in U) on a particle P while $p \notin \mathrm{int}(\nu_e)$—in conflict with axiom C that particles do not leave the cone.

ad (2): A $\mathcal{P}$-geodesic which is timelike at an event must be a particle. (For, there will also then be a 'particle' $\mathcal{P}$-geodesic at the same event with the same tangent; as a geodesic equation is a second-order equation, this means that the two $\mathcal{P}$-geodesics are identical.) From this, the result follows from part (1) of the lemma. □

With this lemma established, EPS then provide a number of definitions:

Definition. *The inverse $\mathcal{g}^{\mu\nu}$ of the conformal metric density is given by*

$$\mathcal{g}^{\mu\nu}\mathcal{g}_{\nu\lambda} = \delta^{\mu}_{\lambda}.$$

Definition. *The components of the conformal connection are given by*

$$K^{\mu}_{\nu\lambda} := \frac{1}{2}\mathcal{g}^{\mu\sigma}\left(\mathcal{g}_{\sigma\nu,\lambda} + \mathcal{g}_{\sigma\lambda,\nu} - \mathcal{g}_{\nu\lambda,\sigma}\right). \tag{1.3}$$

Definition. *The difference between the projective and conformal connection is given by*

$$\Delta^{\mu}_{\nu\lambda} := \Pi^{\mu}_{\nu\lambda} - K^{\mu}_{\nu\lambda},$$

$$\Delta_{\mu\nu\lambda} := \mathcal{g}_{\mu\sigma}\Delta^{\sigma}_{\nu\lambda}.$$

$\Delta^{\mu}_{\nu\lambda}$ can now be shown to fulfil certain conditions. First of all:

Lemma ((a) Properties of $\Delta^{\mu}_{\nu\lambda}$). $\Delta^{\mu}_{\nu\lambda}$ *satisfies* $\Delta^{\mu}_{[\nu\lambda]} = \Delta^{\mu}_{\nu\mu} = 0$.

Proof. For showing $\Delta^{\mu}_{[\nu\lambda]} = 0$, note that we have already seen that $\Pi^{\mu}_{[\nu\lambda]} = 0$, so it suffices to show that $K^{\mu}_{[\nu\lambda]} = 0$. To demonstrate this, consider

$$K^{\mu}_{[\nu\lambda]} = \frac{1}{2}\mathcal{g}^{\mu\sigma}(\mathcal{g}_{\sigma[\nu,\lambda]} + \mathcal{g}_{\sigma[\lambda,\nu]} - \mathcal{g}_{[\nu\lambda],\sigma}).$$

The final term on the RHS here vanishes by the symmetry of $\mathcal{g}_{\mu\nu}$; the first two terms on the RHS cancel; thus, the result follows.

For $\Delta^{\mu}_{\nu\mu} = 0$, consider

$$K^{\mu}_{\nu\mu} = \frac{1}{2}\mathcal{g}^{\mu\sigma}(\mathcal{g}_{\sigma\nu,\mu} + \mathcal{g}_{\sigma\mu,\nu} - \mathcal{g}_{\nu\mu,\sigma}) = 0;$$

since we have already argued above that for a projective connection one can take $\Pi^{\mu}_{\nu\mu} = 0$, the result follows. □

With all of these results established, EPS then prove the following claim, the purpose of which is to learn more regarding the relationship between the projective structure $\mathcal{P}$ and the conformal structure $\mathcal{C}$:

Claim ($\mathcal{P}$-geodesic with tangent $\mathcal{C}$-vector at a point). *Let $x^{\mu}(u)$ describe a $\mathcal{P}$-geodesic such that $\dot{x}^{\mu}(0)$ is a $\mathcal{C}$-null vector. Then, at $u = 0$,*

$$\frac{d}{du}(\mathcal{g}_{\mu\nu}\dot{x}^{\mu}\dot{x}^{\nu}) = -\tilde{\Delta}_{\mu\nu\lambda}\dot{x}^{\mu}\dot{x}^{\nu}\dot{x}^{\lambda},$$

where $\tilde{\Delta}_{\mu\nu\lambda} = \frac{1}{2}\Delta_{\mu\nu\lambda}$.[33]

Proof. One can establish this result by manipulating the geodesic equation as follows:

$$\begin{aligned}
\ddot{x}^{\mu} + \Pi^{\mu}_{\nu\lambda}\dot{x}^{\nu}\dot{x}^{\lambda} &= \lambda\dot{x}^{\mu} \\
\Rightarrow \ddot{x}^{\mu} + \Delta^{\mu}_{\nu\lambda}\dot{x}^{\nu}\dot{x}^{\lambda} + K^{\mu}_{\nu\lambda}\dot{x}^{\nu}\dot{x}^{\lambda} &= \lambda\dot{x}^{\mu} \\
\Rightarrow \ddot{x}^{\mu} + \Delta^{\mu}_{\nu\lambda}\dot{x}^{\nu}\dot{x}^{\lambda} + \frac{1}{2}\mathcal{g}^{\mu\sigma}\left(\partial_{\lambda}\mathcal{g}_{\sigma\nu} + \partial_{\nu}\mathcal{g}_{\sigma\lambda} - \partial_{\sigma}\mathcal{g}_{\nu\lambda}\right)\dot{x}^{\nu}\dot{x}^{\lambda} &= \lambda\dot{x}^{\mu} \\
\Rightarrow \mathcal{g}_{\delta\mu}\ddot{x}^{\mu} + \mathcal{g}_{\delta\mu}\Delta^{\mu}_{\nu\lambda}\dot{x}^{\nu}\dot{x}^{\lambda} + \frac{1}{2}\mathcal{g}_{\delta\mu}\mathcal{g}^{\mu\sigma}\left(\partial_{\lambda}\mathcal{g}_{\sigma\nu} + \partial_{\nu}\mathcal{g}_{\sigma\lambda} - \partial_{\sigma}\mathcal{g}_{\nu\lambda}\right)\dot{x}^{\nu}\dot{x}^{\lambda} &= \lambda\mathcal{g}_{\delta\mu}\dot{x}^{\mu} \\
\Rightarrow \mathcal{g}_{\delta\mu}\ddot{x}^{\mu} + \Delta_{\delta\nu\lambda}\dot{x}^{\nu}\dot{x}^{\lambda} + \frac{1}{2}\left(\partial_{\lambda}\mathcal{g}_{\delta\nu} + \partial_{\nu}\mathcal{g}_{\delta\lambda} - \partial_{\delta}\mathcal{g}_{\nu\sigma}\right)\dot{x}^{\nu}\dot{x}^{\lambda} &= \lambda\mathcal{g}_{\delta\mu}\dot{x}^{\mu} \\
\Rightarrow \mathcal{g}_{\delta\mu}\ddot{x}^{\mu}\dot{x}^{\delta} + \Delta_{\delta\nu\lambda}\dot{x}^{\nu}\dot{x}^{\lambda}\dot{x}^{\delta} + \frac{1}{2}\partial_{\lambda}\mathcal{g}_{\delta\nu}\dot{x}^{\nu}\dot{x}^{\lambda}\dot{x}^{\delta} &= \lambda\mathcal{g}_{\delta\mu}\dot{x}^{\mu}\dot{x}^{\delta} \\
\Rightarrow \frac{d}{du}\left(\mathcal{g}_{\delta\mu}\dot{x}^{\mu}\dot{x}^{\delta}\right)\Big|_{u=0} + \frac{1}{2}\Delta_{\delta\nu\lambda}\dot{x}^{\nu}\dot{x}^{\lambda}\dot{x}^{\delta} &= 0
\end{aligned}$$

where we have used in the penultimate line that $\dot{x}^{\mu}$ is null at $u = 0$ (allowing us for dropping the RHS). □

[33] Note that in the original EPS paper $\tilde{\Delta}_{\mu\nu\lambda} = \Delta_{\mu\nu\lambda}$.

Using this claim, one can then prove the following lemma:[34]

Lemma ((b) Property of $\Delta^{\mu}_{\nu\lambda}$). $\Delta_{(\mu\nu\lambda)} = \mathscr{g}_{(\mu\nu}s_{\lambda)}$ *identically, for some one-form* s_{μ}.

Proof. If the expression $\Delta_{\mu\nu\lambda}\,\dot{x}^{\mu}\dot{x}^{\nu}\dot{x}^{\lambda} = 0$ holds for a non-null vector $\dot{x}^{\mu}$ at $u = 0$, then there is a change from a timelike (spacelike) vector for $u < 0$ to a spacelike (timelike) vector for $u > 0$ (see the proven claim before). This contradicts the first lemma derived from axiom C above, according to which every particle which is a $\mathcal{P}$-geodesic is nowhere spacelike.

So, $\Delta_{\mu\nu\lambda}\,\dot{x}^{\mu}\dot{x}^{\nu}\dot{x}^{\lambda} = 0$ holds for a null vector $\dot{x}^{\mu}$, then $\Delta_{\mu\nu\lambda}\,\dot{x}^{\mu}\dot{x}^{\nu}\dot{x}^{\lambda}$ must be a function f of $\mathscr{g}_{\mu\nu}\dot{x}^{\mu}\dot{x}^{\nu}$. One way to realise this concretely is to set $\Delta_{(\mu\nu\lambda)} = \mathscr{g}_{(\mu\nu}s_{\lambda)}$ for arbitrary one-form s_{λ} such that $\Delta_{\mu\nu\lambda}\,\dot{x}^{\mu}\dot{x}^{\nu}\dot{x}^{\lambda} = (\mathscr{g}_{\mu\nu}\dot{x}^{\mu}\dot{x}^{\nu})s_{\lambda}\dot{x}^{\lambda}$. This way of doing so is unique: (1) the expression is $\Delta_{\mu\nu\lambda}\,\dot{x}^{\mu}\dot{x}^{\nu}\dot{x}^{\lambda} = 0$ is (exactly) a third-order polynomial in $\dot{x}$ (after all, $\Delta^{\mu}_{\nu\lambda} = \Pi^{\mu}_{\nu\lambda} - K^{\mu}_{\nu\lambda}$ and neither the connections $\Pi^{\mu}_{\nu\lambda}$ nor $K^{\mu}_{\nu\lambda}$ are functions of tangent vectors); (2) the factorisation into irreducible polynomials must include $\mathscr{g}_{\mu\nu}\dot{x}^{\mu}\dot{x}^{\nu}$ (as $\mathscr{g}_{\mu\nu}\dot{x}^{\mu}\dot{x}^{\nu}$ encodes two of the zeros, and factorial decomposition of a multivariate polynomial over a field is unique (Lee, 2013, see proposition 2.1 and subsequent comments)); (3) the remaining part has to be proportional to $\dot{x}^{\lambda}$ to keep the exact third order of the polynomial. □

EPS are then in position to establish the first results on the relation between $\mathcal{P}$ and $\mathcal{C}$:

Claim (Relation between $\mathcal{P}$ and $\mathcal{C}$). *On any* $\mathcal{P}$*-geodesic* $x^{\mu}(u)$*, the following relation obtains:*

$$\frac{d}{du}(\mathscr{g}_{\mu\nu}\dot{x}^{\mu}\dot{x}^{\nu}) = (\mathscr{g}_{\mu\nu}\dot{x}^{\mu}\dot{x}^{\nu})(2\lambda - s_{\lambda}\dot{x}^{\lambda}).$$

Proof. Consider the penultimate line of the proof of the above claim. (We cannot use the last line of the proof in claim ($\mathcal{P}$-geodesic with tangent $\mathcal{C}$-vector at a point) as one is considering a generic $\mathcal{P}$-geodesic here, not one necessarily tangent to a $\mathcal{C}$-vector.) Inserting the identity from the above lemma (while also re-scaling λ), one obtains the desired result. □

[34] Cf. Kriele (1999), Lemma 3.3.2 (p. 161).

Corollary (Relation between $\mathcal{P}$ and $\mathcal{C}$). *A $\mathcal{P}$-geodesic that is timelike, spacelike, or null respectively, with respect to $\mathcal{C}$ at one of its events, has the same orientation everywhere.*

Proof. Let $f(u) := g_{\mu\nu}\dot{x}^{\mu}\dot{x}^{\nu}$. If $\dot{x}^{\mu}$ is a null vector for some u_0, then since $\dot{f} = 0$ in light of the previous claim (Relation between $\mathcal{P}$ and $\mathcal{C}$), $\dot{x}^{\mu}$ is null for *all* u (by the above). If $\dot{x}^{\mu}$ is timelike or spacelike at some parameter time u_0, proceed via a case distinction: on the kernel of f, $\dot{f} = 0$ so that $f =$ const.; otherwise we can write $\dot{f}/f = 2\lambda - s_{\lambda}\dot{x}^{\lambda}$ and logarithmic integration gives that

$$\log f(u_1) - \log f(u_0) = \int_{u_0}^{u_1} du(2\lambda - s_{\lambda}\dot{x}^{\lambda}),$$

and ultimately that

$$f(u_1) = f(u_0) \cdot \exp(\ldots).$$

As $\exp(\ldots) > 0$, the sign of $f(u_1)$ for any u_1 is the same as for $f(u_0)$. □

One can now define:[35]

Definition (EPS-compatibility between projective and conformal structure). *A projective structure and a conformal structure are* EPS-compatible *iff the null geodesics of the conformal structure are also projective geodesics (but not necessarily vice versa). (See Trautman (2012).)*

Before showing EPS-compatibility between $\mathcal{P}$ and $\mathcal{C}$, the following lemma is established:

Lemma. *Each event p on v_e sufficiently close to e can be approximated arbitrarily closely by events q situated on particles through e.*

Proof. Axiom C states that for any event e there is a neighbourhood such that all events within that neighbourhood *that are connected to e via particles* will lie inside v_e. If p is sufficiently close to e, then it is sufficiently close to events q inside v_e that are connected to e via particles. □

[35] Note that Scholz (2020) uses the terminology ‘null line’ here rather than ‘null geodesic’.

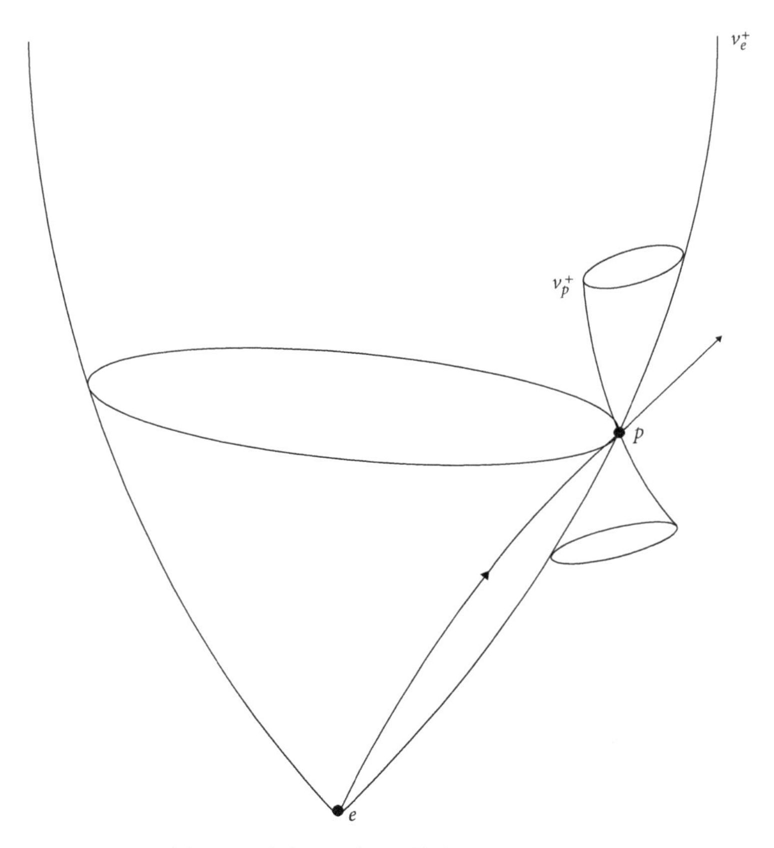

Figure 9 in Ehlers et al. (2012 [1972]) (own drawing)

Proposition (EPS-compatibility). *$\mathcal{C}$ and $\mathcal{P}$ are EPS-compatible.*

Proof. Let $p \in \nu_e^+, p \neq e$. (The same argument holds for ν_e^-). Let $(q_n)_{n\in\mathbb{N}}$ be a sequence of events within a $\mathcal{P}$-convex neighbourhood of e such that (i) $q_n \to p$ where '$\to$' is to denote topological convergence here, and (ii) each sequence element q_n together with e defines the sequence element P_n in the series of geodesics (particles) $(P_n)_{n\in\mathbb{N}}$.[36] That (i) can be stipulated like this, follows from the lemma just proven based on axiom C. Notably, each particle P_n can be taken to obey the (projective) geodesic equation $\ddot{x} + \Pi^\lambda_{\mu\nu}\dot{x}^\mu\dot{x}^\nu = \lambda_n\dot{x}$ parameterised in such a way that, for $\lambda_n = 0$, P_n goes through e, and, for $\lambda_n = 1$, P_n goes through q_n. In particular, in the limit $q_n \to q$, the corresponding geodesic

[36] A sequence of points $(s_n)_{n\in\mathbb{N}}$ in the topological space X is said to converge topologically to s if for each neighbourhood U of s there exists a natural number N such that $\forall K \geq N$: $s_K \in U$.

$P \leftarrow P_n$ goes through e, and for $\lambda = 1$ the geodesic goes through $p \in \nu_e$; and the sequence $\{T_n\}_{n \in \mathbb{N}}$ where T_n is the tangent vector of P_n at e converges to the tangent vector T of that geodesic P. (All of this assumes that q_n, and p lie in a sufficiently small $\mathcal{P}$-convex coordinate neighbourhood of e.)

For $T_n \rightarrow T$ with T_n non-spacelike, T is either timelike or null: for any tangent vector S, the map $S^\mu \mapsto \mathscr{g}_{\mu\nu} S^\mu S^\nu$ is a continuous function, i.e. convergence of a sequence S^μ implies the convergence of $\mathscr{g}_{\mu\nu} S^\mu S^\nu$. As for all T_n, $\mathscr{g}_{\mu\nu} T_n^\mu T_n^\nu < 0$, thereby $\mathscr{g}_{\mu\nu} T^\mu T^\nu \leq 0$. At the same time, we can exclude that T is timelike: for if it were timelike, then P—coming from e—would have had to intersect ν_e^+ and would therefore have to be spacelike at p (see Figure 9), which is disallowed.

We now want to show that, between passing e and p, P can only contain events on ν_e, i.e. $P \subset \nu_e$. First, $q \in P$ cannot be strictly outside[37] of the cone ν_e between passing e and p because then at least at some point $\mathscr{g}_{\mu\nu} T_n^\mu T_n^\nu > 0$, in contradiction with P being timelike or null at any point by construction. Secondly, assume $q \in P$ was a point strictly inside the cone ν_e upon passing from e to p. Then: $q \in \nu_q^+$; but at least locally, $\nu_q^+ \cap \nu_e^+ = \emptyset$. But then a sufficiently close p could not be reached by P any more. Thus, q must be on ν_e.

To summarise: (1) the tangents along P are null between e and p (what we said for the tangent vector at p holds without any restriction for the tangents at the points to P between e and p too). P is thereby a projective null geodesic. (2) Between e and p, P is contained in ν_e; for this region then, P is a $\mathfrak{C}$-null geodesic. Given that being a geodesic is a local property and that the considerations above hold around any other event e, it follows that for every null geodesic a projective null geodesic can be constructed that coincides with it. □

It is worth pointing out that EPS claim in their main text (at the bottom of p. 80) to show that 'projective null geodesics and conformal null geodesics

[37] Points inside of the cone ν_e are associated with a negative world function value, points outside of ν_e with a positive world function where the world function is defined, up to positive re-scaling, using the conformal structure. Concretely, the world function relative to the base point e and its neighbouring point p is defined as

$$\sigma(e,p) = \frac{1}{2}(\lambda_1 - \lambda_0) \int_{\lambda_0}^{\lambda_1} g_{\mu\nu}(z(\lambda)) t^\mu t^\nu d\lambda, \tag{1.4}$$

with $t^\mu := \frac{dz^\mu}{d\lambda}$ tangent to the geodesic, and $g_{\mu\nu}$ a representative of the conformal structure $\mathfrak{C}$.

are identical'. Note that they only show that every $\mathcal{C}$-null geodesic is a projective (null) geodesic (EPS-compatibility)—and not its converse—which is, however, sufficient for everything that follows below.

In any case, EPS-compatibility between $\mathcal{C}$ and $\mathcal{P}$ has two important corollaries:

Corollary (Algebraic condition for EPS-compatibility). *The following condition holds for $\mathcal{P}$ and $\mathcal{C}$:*

$$\Delta^{\mu}_{\nu\lambda} = 5q^{\mu}\mathcal{g}_{\nu\lambda} - 2\delta^{\mu}_{(\nu}q_{\lambda)} \tag{1.5}$$

where q^{λ} is a vector and q_{λ} the corresponding one-form.

We know that $\mathcal{P}$ and $\mathcal{C}$ are EPS-compatible, so we can use this result in what follows. We begin by presenting the proof of this corollary as presented by EPS (after presenting and proving a helpful lemma); we then present an alternative proof by Matveev and Scholz (2020).

Lemma (Properties of $L_{\mu\nu\lambda}$). *Define* $L_{\mu\nu\lambda} := \frac{4}{3}\Delta_{[\mu\nu]\lambda} - p_{[\mu}\mathcal{g}_{\nu]\lambda}$ *with* $p_{\mu} := -\frac{8}{9}\Delta^{\sigma}_{[\mu\sigma]}$. *Then:*

1. $L_{(\mu\nu)\lambda} = 0$.
2. $L_{[\mu\nu\lambda]} = 0$.
3. $L^{\mu}_{\nu\mu} = 0$.
4. $\Delta_{\mu\nu\lambda} = \Delta_{(\mu\nu\lambda)} + \frac{1}{2}(p_{\mu}\mathcal{g}_{\nu\lambda} - \mathcal{g}_{\mu(\nu}p_{\lambda)}) + L_{\mu(\nu\lambda)}$.

Proof. It is helpful to expand $L_{\mu\nu\lambda}$ according to its definition:

$$\begin{aligned} L_{\mu\nu\lambda} &= \frac{4}{3}\Delta_{[\mu\nu]\lambda} - p_{[\mu}\mathcal{g}_{\nu]\lambda} \\ &= \frac{2}{3}\left(\Delta_{\mu\nu\lambda} - \Delta_{\nu\mu\lambda}\right) - \frac{1}{2}\left(p_{\mu}\mathcal{g}_{\nu\lambda} - p_{\nu}\mathcal{g}_{\mu\lambda}\right) \\ &= \frac{2}{3}\left(\Delta_{\mu\nu\lambda} - \Delta_{\nu\mu\lambda}\right) + \frac{2}{9}\left(\Delta^{\rho}_{\mu\rho}\mathcal{g}_{\nu\lambda} - \Delta^{\rho}_{\rho\mu}\mathcal{g}_{\nu\lambda} - \Delta^{\rho}_{\nu\rho}\mathcal{g}_{\mu\lambda} + \Delta^{\rho}_{\rho\nu}\mathcal{g}_{\mu\lambda}\right). \end{aligned}$$

Notably, we will only need the last identity for showing (3); so (1), (2), and (4) hold for any p.[38]

[38] This will become important again when we are using the lemma below, cf. footnote 39.

ad (1): The first object is zero due to the definition of $L_{\mu\nu\lambda}$ as anti-symmetric in μ and ν. (This can also be explicitly seen from the expansion just given.)

ad (2): We know from lemma ((a) Properties of $\Delta^{\mu}_{\nu\lambda}$) that $\Delta_{\mu\nu\lambda} = \Delta_{\mu(\nu\lambda)}$. Thus, upon symmetrisation the first two terms on the RHS of (1.7) cancel each other out, i.e., $\Delta_{[\mu\nu\lambda]} - \Delta_{[\nu\mu\lambda]} = \Delta_{[\mu\nu\lambda]} - \Delta_{[\nu\lambda\mu]} = \Delta_{[\mu\nu\lambda]} - \Delta_{[\mu\nu\lambda]} = 0$. Furthermore, due to the symmetry of $\mathscr{g}_{\mu\lambda} = \mathscr{g}_{(\mu\lambda)}$ (cf. Lemma (Properties of g), (c)), $p_{[\mu}\mathscr{g}_{\nu\lambda]} - p_{[\nu}\mathscr{g}_{\mu\lambda]} = p_{[\mu}\mathscr{g}_{\nu\lambda]} - p_{[\nu}\mathscr{g}_{\lambda\mu]} = p_{[\mu}\mathscr{g}_{\nu\lambda]} - p_{[\mu}\mathscr{g}_{\nu\lambda]} = 0$; so the result follows.

ad (3): That the third object is zero follows from direct calculation: $L^{\mu}{}_{\nu\mu} = \frac{2}{3}\left(\Delta^{\rho}{}_{\nu\rho} - \Delta_{\nu}{}^{\rho}{}_{\rho}\right) + \frac{2}{9}\left(\Delta_{\nu\rho}{}^{\rho} - \Delta_{\rho\nu}{}^{\rho} - 4\Delta_{\nu\rho}{}^{\rho} + 4\Delta_{\rho\nu}{}^{\rho}\right) = 0$.

ad (4): Given the definition of $L_{\mu\nu\rho}$, $L_{\mu(\nu\rho)} = \frac{2}{3}(\Delta_{[\mu\nu]\rho} + \Delta_{[\mu\rho]\nu}) - \frac{1}{2}(p_{\mu}\mathscr{g}_{\nu\rho} - p_{(\nu}\mathscr{g}_{\rho)\mu})$. Inserting this into the equality to demonstrate, what remains to be shown is that $\Delta_{\mu\nu\rho} \overset{!}{=} \Delta_{(\mu\nu\rho)} + \frac{2}{3}(\Delta_{[\mu\nu]\rho} + \Delta_{[\mu\rho]\nu})$. Going from right to left:

$$
\begin{aligned}
\Delta_{(\mu\nu\rho)} + \frac{2}{3}(\Delta_{[\mu\nu]\rho} + \Delta_{[\mu\rho]\nu}) &= \frac{1}{6}(\Delta_{\mu\nu\rho} + \ldots) + \frac{1}{3}(\Delta_{\mu\nu\rho} - \Delta_{\nu\mu\rho} + \Delta_{\mu\rho\nu} - \Delta_{\rho\mu\nu}) \\
&= \frac{1}{2}\Delta_{\mu\nu\rho} + \frac{1}{6}\Delta_{\nu\rho\mu} - \frac{1}{6}\Delta_{\rho\mu\nu} - \frac{1}{6}\Delta_{\nu\mu\rho} + \frac{1}{6}\Delta_{\rho\nu\mu} + \frac{1}{2}\Delta_{\mu\rho\nu} \\
&= \frac{1}{2}\Delta_{\mu\nu\rho} + \frac{1}{12}\Delta_{\nu[\rho\mu]} + \frac{1}{12}\Delta_{\rho[\nu\mu]} + \frac{1}{2}\Delta_{\mu\rho\nu} \\
&= \Delta_{\mu\nu\rho},
\end{aligned}
$$

where the last steps follows from that $\Delta_{\mu[\nu\rho]} = 0$ (and thus also $\Delta_{\mu\nu\rho} = \Delta_{\mu(\nu\rho)}$). □

The original proof by EPS is heavy in algebraic manipulations. The alternative proof due to Matveev and Scholz (2020) is arguably a more straightforward alternative (in fact, they first establish a more general result where $q^{\mu} \rightarrow q'^{\mu}$ such that q'^{μ} is not necessarily $g^{\mu\nu}q_{\mu}$).

Proof. (Algebraic conditions for EPS-compatibility, EPS version)

$\mathscr{g}_{\mu\nu}\dot{x}^{\mu}\dot{x}^{\nu} = 0$ and $\ddot{x}^{\mu} + K^{\mu}_{\nu\lambda}\dot{x}^{\mu}\dot{x}^{\nu} = \nu\dot{x}^{\mu}$ together characterise $\mathfrak{C}$-null geodesics; as these are, however, also projective geodesics, they may also be described as $\ddot{x}^{\mu} + \Pi^{\mu}_{\nu\lambda}\dot{x}^{\nu}\dot{x}^{\lambda} = \lambda\dot{x}^{\mu}$. Subtracting the second from the third equation gives $\Delta^{\mu}_{\nu\lambda}\dot{x}^{\nu}\dot{x}^{\lambda} = (\lambda - \nu)\dot{x}^{\mu}$. Using the second part of Lemma ((a) Properties of $\Delta^{\mu}_{\nu\lambda}$), i.e., $\Delta_{\mu\nu\lambda} = \Delta_{(\mu\nu\lambda)} + \frac{1}{2}(p_{\mu}\mathscr{g}_{\nu\lambda} - \mathscr{g}_{\mu(\nu}p_{\lambda)}) + L_{\mu(\nu\lambda)}$, together with $\Delta^{\mu}_{[\nu\lambda]} = \Delta^{\mu}_{\nu\mu} = 0$ and $L_{(\mu\nu)\lambda} = L_{[\mu\nu\lambda]} = L^{\mu}_{\nu\mu} = 0$ gives $\Delta^{\mu}_{\nu\lambda} = L^{\mu}_{(\nu\lambda)} + 5q^{\mu}\mathscr{g}_{\nu\lambda} - 2\delta^{\mu}{}_{(\nu}q_{\lambda)}$. Then, we have:

$$\begin{aligned}
\Delta^{\mu}_{\nu\lambda} &= \mathcal{g}^{\mu\alpha}\Delta_{(\alpha\nu\lambda)} + \frac{1}{2}(p^{\mu}\mathcal{g}_{\nu\lambda} - \mathcal{g}^{\mu\alpha}\mathcal{g}_{\alpha(\nu}p_{\lambda)}) + L^{\mu}_{(\nu\lambda)} \\
&= \mathcal{g}^{\mu\alpha}\mathcal{g}_{(\alpha\nu}s_{\lambda)} + \frac{1}{2}p^{\mu}\mathcal{g}_{\nu\lambda} - \delta^{\mu}_{(\nu}p_{\lambda)} + L^{\mu}_{(\nu\lambda)} \\
&= \frac{1}{6}\mathcal{g}^{\mu\alpha}\mathcal{g}_{\alpha\nu}s_{\lambda} + \frac{1}{6}\mathcal{g}^{\mu\alpha}\mathcal{g}_{\nu\lambda}s_{\alpha} + \frac{1}{6}\mathcal{g}^{\mu\alpha}\mathcal{g}_{\lambda\alpha}s_{\nu} + \frac{1}{6}\mathcal{g}^{\mu\alpha}\mathcal{g}_{\nu\alpha}s_{\lambda} \\
&\qquad + \frac{1}{6}\mathcal{g}^{\mu\alpha}\mathcal{g}_{\alpha\lambda}s_{\nu} + \frac{1}{6}\mathcal{g}^{\mu\alpha}\mathcal{g}_{\lambda\nu}s_{\alpha} + \frac{1}{2}p^{\mu}\mathcal{g}_{\nu\lambda}\delta^{\mu}_{(\nu}p_{\lambda)} + L^{\mu}_{(\nu\lambda)} \\
&= \frac{1}{6}\delta^{\mu}_{\nu}s_{\lambda} + \frac{1}{6}\mathcal{g}_{\nu\lambda}s^{\mu} + \frac{1}{6}\delta^{\mu}{}_{\lambda}s_{\nu} + \frac{1}{6}\delta^{\mu}_{\nu}s_{\lambda} \\
&\qquad + \frac{1}{6}\delta^{\mu}_{\lambda}s_{\nu} + \frac{1}{6}\mathcal{g}_{\lambda\nu}s^{\mu} + \frac{1}{2}p^{\mu}\mathcal{g}_{\nu\lambda} - \delta^{\mu}_{(\nu}p_{\lambda)} + L^{\mu}_{(\nu\lambda)} \\
&= \frac{2}{3}\delta^{\mu}{}_{(\nu}s_{\lambda)} - \delta^{\mu}_{(\nu}p_{\lambda)} + \frac{1}{3}\mathcal{g}_{\nu\lambda}s^{\mu} + \frac{1}{2}\mathcal{g}_{\nu\lambda}p^{\mu} + L^{\mu}_{(\nu\lambda)}.
\end{aligned}$$

Now, letting $p_{\mu} \equiv s_{\mu}$[39] and replacing s_{μ} by $q_{\mu} := \frac{s_{\mu}}{6}$, we obtain

$$\Delta^{\mu}_{\nu\lambda} = L^{\mu}_{(\nu\lambda)} + 5q^{\mu}\mathcal{g}_{\nu\lambda} - 2\delta^{\mu}_{(\nu}q_{\lambda)}. \tag{1.6}$$

What remains to be done is to show that $L^{\mu}_{(\nu\lambda)} = 0$. From (1.6) and the previous identity for $\Delta^{\mu}_{\nu\lambda}$, we obtain, for null curves with tangent T^{μ},

$$L^{\mu}_{(\nu\lambda)}T^{\nu}T^{\lambda} - 2\delta^{\mu}_{(\nu}q_{\lambda)}T^{\nu}T^{\lambda} = (\lambda - \nu)T^{\mu}.$$

As $2\delta^{\mu}{}_{(\nu}q_{\lambda)}T^{\nu}T^{\lambda} = 2q_{\lambda}T^{\lambda}T^{\mu}$, we have

$$L^{\mu}_{\nu\lambda}T^{\nu}T^{\lambda} = L^{\mu}_{(\nu\lambda)}T^{\nu}T^{\lambda} \propto T^{\mu}.$$

As, again, the latter equation holds for any null vector T^{μ}, it follows that $L^{\mu}_{(\nu\lambda)} = 0$, so the stated result follows. □

Proof. (Algebraic conditions for EPS-compatibility, Matveev-Scholz version)[40]

[39] Use here that $L_{\mu\nu\lambda}$ (as defined in lemma (Properties of $L_{\mu\nu\lambda}$)) can be seen to hold for any p—unless property (3) is invoked, which requires further specification of p—as originally done in the lemma. However, no recourse to property (3) is needed here or in the following.

[40] We are basically just restating the proof by Matveev and Scholz (2020); we, however, follow EPS' notation for conformal and projective structure and use a conformal metric tensor density $\mathcal{g}$ rather than a metric tensor g which is a representative of the equivalence class picked out by $\mathcal{g}$.

For the lightlike geodesic γ,

$$\nabla^{\mathscr{g}}_{\dot{\gamma}} \dot{\gamma} = 0$$
$$\nabla^{\Pi}_{\dot{\gamma}} \dot{\gamma} = \beta(\gamma, \dot{\gamma}),$$

where $\nabla^{\mathscr{g}}$ is the coordinate-free conformal connection with coordinate components $K^{\mu}_{\mu\lambda}$, and ∇^{Π} is the coordinate-free projective connection with coordinate components $\Pi^{\mu}_{\mu\lambda}$. Then, subtracting gives

$$\underbrace{(\Pi^{\mu}_{\nu\lambda} - K^{\mu}_{\nu\lambda})\,\dot{\gamma}^{\lambda}}_{=\Delta^{\mu}_{\nu\lambda}\dot{\gamma}^{\nu}} = \beta(\gamma, \dot{\gamma})\dot{\gamma}^{\mu}. \tag{1.7}$$

where, recall, $\Pi^{\mu}_{\nu\lambda}$ are the components of the projective connection under consideration, and $K^{\mu}_{\nu\lambda}$ are the components of the conformal connection under consideration.

For any vector v^{μ}, consider now the object

$$\Delta^{\mu}_{\nu\lambda} v^{\nu} v^{\lambda} v^{\rho} - \Delta^{\rho}_{\nu\lambda} v^{\nu} v^{\lambda} v^{\mu}. \tag{1.8}$$

We see immediately that this expression must be zero for v^{μ} light-like (as then the object becomes $\beta v^{\mu} v^{\beta} - \beta v^{\beta} v^{\mu} = 0$). Given that $g(v, v)$ is an irreducible quadric, the expression equals $\mathscr{g}(v, v)\omega^{\mu\rho}{}_{\sigma} v^{\sigma}$ for some $\omega^{\mu\rho}{}_{\sigma}$ antisymmetric with respect to the indices μ, ρ. (Equation (1.8) can be thought of as setting up a multivariate polynomial over the field $\mathbb{R}$; notably, the factorisation of polynomials over fields is unique since fields themselves are (trivially) factorial (Lee, 2013, see proposition 2.1 and subsequent comments).)

At the same time, (1.8) is such that for any two one-forms σ_{μ}, ρ_{μ} with $\sigma_{\mu} v^{\mu} = 0, \rho_{\mu} v^{\mu} = 0$, contraction of (1.8) with σ_{μ}, ρ_{μ} yields zero. Therefore, at any v and for any two one-forms σ_{μ}, ρ_{μ} with $\sigma_{\mu} v^{\mu} = 0, \rho_{\mu} v^{\mu} = 0$:

$$\sigma_{\mu} \rho_{\rho} \omega^{\mu\rho}{}_{\sigma} v^{\sigma} = 0. \tag{1.9}$$

Given that $\omega^{\mu\rho}{}_{\sigma} v^{\sigma}$ is a second-rank tensor in μ and ρ, (1.9) means that it must decompose into $\eta^{\mu} v^{\rho} + v^{\mu} \zeta^{\rho}$ (for a fixed v, other terms than those 'proportional to v' would not become zero by contraction with arbitrary σ_{μ}, ρ_{μ} for which $\sigma_{\mu} v^{\mu} = 0, \rho_{\mu} v^{\mu} = 0$). Due to the anti-symmetry in μ and ρ, it follows that $v^{\mu} \phi^{\rho} - \phi^{\mu} v^{\rho}$ for some ϕ.

Based on the above, we now know that

$$\Delta^{\mu}_{\nu\lambda}v^{\nu}v^{\lambda}v^{\rho} - \Delta^{\rho}_{\nu\lambda}v^{\nu}v^{\lambda}v^{\mu} = (\phi^{\mu}v^{\rho} - v^{\rho}\phi^{\mu})\, \mathscr{g}_{\alpha\beta}v^{\alpha}v^{\beta} \tag{1.10}$$

One can treat this as a set of linear inhomogeneous differential equations, the corresponding homogeneous system for which is

$$\Delta^{\mu}_{\nu\lambda}v^{\nu}v^{\lambda}v^{\rho} - \Delta^{\rho}_{\nu\lambda}v^{\nu}v^{\lambda}v^{\mu} = 0. \tag{1.11}$$

Then, view (1.10) as a linear inhomogeneous equation system for Δ (with ϕ known), that is fulfilled for any v; by finding the solution, we will find the desired identity. The solution for D will be a linear combination of the general solution to the homogeneous system associated with (1.11) together with a specific solution to an inhomogeneous system. Now, finding the solution space for the homogeneous system is equivalent to the condition that for every v^{μ},

$$\Delta^{\mu}_{\nu\lambda}v^{\nu}v^{\lambda} \propto v^{\mu}. \tag{1.12}$$

But this only means that the solution is the space of free trace tensors so: $\Delta^{\mu}_{\nu\lambda} = \delta^{\mu}_{\nu}\eta_{\lambda}+\delta^{\mu}_{\lambda}\eta_{\nu}$; a specific solution to the inhomogeneous solution is $\Delta^{\mu}_{\nu\lambda} = -\phi^{\mu}\,\mathscr{g}_{\nu\lambda}$. Putting the solutions together gives

$$\Delta^{\mu}_{\nu\lambda} = \delta^{\mu}_{\nu}\eta_{\lambda} + \delta^{\mu}_{\lambda}\eta_{\nu} - \phi^{\mu}\,\mathscr{g}_{\nu\lambda}. \tag{1.13}$$

For $\phi^{\mu} =: -5q^{\mu}$ and $\eta_{\mu} =: -q'_{\mu}$, we get nearly the identity sought by EPS (note that q and q' are distinct here):

$$\Delta^{\mu}_{\nu\lambda}(q, q') = -2\delta^{\mu}_{(\nu}q_{\lambda)} + 5q'^{\mu}\,\mathscr{g}_{\nu\lambda}. \tag{1.14}$$

Suppressing now indices for clarity: to arrive at the identity sought by EPS, think of the projective structure Π we started with, given now by $\Pi = K + \Delta(q, q')$. Reparameterising geodesics has the algebraic consequence of adding to $\Delta(q, q')$ a symmetric, trace-free tensor of form $\delta^{\mu}_{(\nu}\psi_{\lambda)}$ for some one-form ψ_{λ}; this projective transformation leaves the projective coefficients Π invariant, and can be used to establish that one can write $\Pi = K + \Delta(q, q)$.[41] □

Corollary. *The set of all particles is identical to the set of all $\mathcal{C}$-timelike $\mathcal{P}$-geodesics.*

[41] In fact, one might directly establish by such a projective equivalence transformation the Weyl connection of the claim (Existence of a Weyl connection (Weyl space formulation)), discussed below.

Proof. We already know that particles are $\mathcal{P}$-geodesics, so what is at stake is whether they are always $\mathcal{C}$-timelike. If a particle were at any point p a $\mathcal{C}$-null geodesic, then it would, in light of the claim previously proven, have to be a $\mathcal{P}$-null geodesic at p (as implied by EPS compatibility), and, in light of corollary (relation between $\mathcal{P}$ and $\mathcal{C}$), then everywhere. But particles are not just $\mathcal{C}$-null given that a particle is, by definition, a geodesic that is timelike at *some* event (see bottom of p. 77 in EPS). This then also rules out that a particle could be $\mathcal{C}$-spacelike: it would have to be $\mathcal{C}$-timelike at some point and thus, by continuity, $\mathcal{C}$-null at some point. Thus, $\mathcal{P}$-geodesics must be $\mathcal{C}$-timelike geodesics everywhere. □

These results on EPS-compatibility give rise to what we call an *EPS-Weyl space*:

Definition (EPS-Weyl space). *A manifold M together with a EPS-compatible pair of conformal structure $\mathcal{C}$ and projective structure $\mathcal{P}$ is called an* EPS-Weyl space $(M, \mathcal{C}, \mathcal{P})$.

We can also define, following Scholz (2020), the notions of *Weyl structure* and *Weyl manifold*—and, relatedly, the notions of *Weyl connection* and *Weyl metric*:[42]

Definition (Weyl structure and Weyl manifold (Scholz 2020, definition 1)).

1. *A (pseudo-Riemannian)* Weyl structure *is given by the triple $(M, \mathcal{C}, \nabla)$ where M is a differentiable manifold, $\mathcal{C} = [g_{\mu\nu}]$ is a conformal class of pseudo-Riemannian metrics $g_{\mu\nu}$ on M (picked out by $g_{\mu\nu} \mapsto \tilde{g}_{\mu\nu} = \Omega^2 g_{\mu\nu}$), and ∇ is the covariant derivative of a torsion-free affine connection Γ (called* Weyl connection*), constrained by the* Weyl compatibility condition *that for any $g_{\mu\nu} \in \mathcal{C}$ there is a differential 1-form φ_μ such that* $\nabla_\lambda g_{\mu\nu} + 2\varphi_\lambda g_{\mu\nu} = 0$.
2. *A* Weyl manifold $(M, [(g_{\mu\nu}, \varphi_\mu)])$ *is a differentiable manifold M endowed with a* Weyl metric *defined by an equivalence class of pairs $(g_{\mu\nu}, \varphi_\mu)$, where $g_{\mu\nu}$ is a pseudo-Riemannian metric on M and φ_μ is a (real valued) differential 1-form on M. Equivalence is defined by conformal rescaling $g_{\mu\nu} \mapsto \tilde{g}_{\mu\nu} = \Omega^2 g_{\mu\nu}$ and the corresponding transformation* $\varphi_\mu \mapsto \varphi_\mu - d_\mu \ln \Omega$.

[42] Note that the following definitions follow Scholz (2020) closely, with only minor modifications to terminology.

Note that 'Weyl manifold' is equivalent to 'Weyl structure' (Scholz, 2020): The Weyl metric induces a Weylian connection, which is an equivalence class of derivative operators each compatible with the (conformally-related) metrics in the class of pairs $(g_{\mu\nu}, \varphi_\mu)$; these derivative operators will satisfy the equation $\nabla_\lambda g_{\mu\nu} + 2\varphi_\lambda g_{\mu\nu} = 0$ which is characteristic of a Weyl structure. On the other hand, from the Weyl structure one can obtain the classes of pairs $(g_{\mu\nu}, \varphi_\mu)$ characteristic of a Weyl manifold, for $\nabla_\lambda g_{\mu\nu} + 2\varphi_\lambda g_{\mu\nu} = 0$ is invariant under the transformations $g_{\mu\nu} \mapsto \tilde{g}_{\mu\nu} = \Omega^2 g_{\mu\nu}$ and $\varphi_\mu \mapsto \tilde{\varphi}_\mu = \varphi_\mu - d_\mu \ln \Omega$.

Note in this context that the family of φ_μ defining a Weyl structure/Weyl manifold differ only by exact forms; the common exterior derivative is called distant curvature (or, equivalently, 'length curvature') ('Streckenkrümmung'), denoted by $F_{\mu\nu}$. That is, $F_{\mu\nu}$ is given by $F_{\mu\nu} = d_\mu \varphi_\nu = d_\mu \tilde{\varphi}_\nu$.[43]

EPS' argument for the claim that an EPS-Weyl space in fact gives rise to a Weyl structure/Weyl manifold is highly schematic and non-rigorous; we thus base our following presentation of how EPS establish a Weyl structure upon the excellent article by Matveev and Scholz (2020).[44]

Claim (Existence of a Weyl connection (Weyl space formulation)). *The expression*

$$\Gamma^\mu_{\nu\lambda} = K^\mu_{\nu\lambda} + 5q^\mu \mathscr{g}_{\nu\lambda} - 10\delta^\mu_{(\nu} q_{\lambda)}$$

defines a Weyl connection on the EPS-Weyl space $(M, \mathcal{C}, \mathcal{P})$ where q_λ, $q^\mu = \mathscr{g}^{\mu\lambda} q_\lambda$.

Proof. That Γ is a Weyl connection is easily checked from its transformation properties. The claim then follows immediately from the fact that the projective connection as determined by the corollary (Algebraic condition for EPS-compatibility), namely $\Pi^\mu_{\nu\lambda} = K^\mu_{\nu\lambda} + \Delta^\mu_{\nu\lambda}$, is related by a projective transformation to $\Gamma^\mu_{\nu\lambda} = K^\mu_{\nu\lambda} + 5q^\mu \mathscr{g}_{\nu\lambda} - 10\delta^\mu_{(\nu} q_{\lambda)}$ ($\Gamma^\mu_{\nu\lambda}$ and $\Pi^\mu_{\nu\lambda}$ just differ by some symmetric term on the RHS of the form $\delta^\mu_{(\nu} \psi_{\lambda)}$ for some one-form ψ_μ). □

[43] Strictly speaking, the distant curvature is only well-motivated along these lines if exact forms are closed. A sufficient condition which will be required from now for neighbourhoods is simply-connectedness.

[44] For a concise pedagogical introduction to Weyl geometry—albeit with a slightly different (but nevertheless illuminating) terminology to that used Matveev and Scholz (2020)—see Folland (1970).

Given the equivalence of Weyl structure and Weyl manifold, this also defines a Weyl metric.

Theorem (Uniqueness of the Weyl metric ('Weyl's theorem'), cited after Scholz (2020)) *A Weyl metric is uniquely determined by its projective and conformal structures.*

Proof. That Γ is a Weyl connection is easily checked from its transformation properties. See Weyl (1921); and our section 2.1.1 for a sketch of the proof. □

We have thus obtained a unique Weyl structure from $\mathcal{P}$ and $\mathcal{C}$, characterisable through the triple $(M, \mathcal{C}, \mathcal{A})$ where $\mathcal{A}$ is the affine structure associated to the Weyl connection (recall that a Weyl connection is affine structure; it identifies a set of parameterised geodesics). There has been an interesting recent discussion—promulgated in the critical introductory comments of Trautman (2012) to the 2012 reprint of EPS in *General Relativity and Gravitation* as a 'golden oldie'—as to whether the EPS condition of compatibility between conformal and projective structure is sufficient (and not just necessary) to establish the existence of a Weyl connection (see in particular Matveev and Trautman (2014)). Ultimately, as rectified by Matveev and Scholz (2020), the compatibility condition stated by EPS is indeed sufficient—we refer readers to that paper for the details. Second, an insightful account of the history of Weyl's theorem and its subsequent appearance in the EPS framework is given by Scholz (2020).

1.6 From Weyl manifold to Lorentzian manifold

EPS can give the following characterisation of a Lorentzian manifold in terms of a Weyl manifold. A Weyl manifold is Lorentzian if and only if one of the following criteria is fulfilled:

Equidistance criterion: 'the proper times t, t' on two arbitrary, infinitesimally close, freely falling particles P, P' are linearly related (to first order in the distance) by Einstein-simultaneity; i.e., whenever $p_1, p_2, \ldots$ is an equidistant sequence of events on P (ticking of a clock) and $q_1, q_2, \ldots$ is the sequence of events on P' that are Einstein-simultaneous with $p_1, p_2, \ldots$ respectively, then $q_1, q_2, \ldots$ is (approximately) an equidistant sequence on P' ...' (p. 69)

Congruence criterion (no second-clock effect): 'consider parallel transport of a vector V_p from a point p to a point q along two different curves P, P'. The resulting vectors, at q, V_q and V'_q, will be different, in general ... If and only if V_q and V'_q are congruent [have the same length] for all such figures, the Weyl geometry considered is in fact, [pseudo-]Riemannian.' (p. 69)

In order to prove that a Weyl manifold is Lorentzian if and only if one of these criteria is fulfilled, one makes recourse to two lemmas:

Lemma (Curvature decomposition in Weyl structure). *The curvature tensor $R^{\lambda}_{\mu cd}$ associated with a Weyl connection decomposes uniquely according to*

$$R^{\lambda}_{\mu\nu\rho} = \hat{R}^{\lambda}_{\mu\nu\rho} + \frac{1}{2}\delta^{\lambda}_{\mu}F_{\nu\rho}, \tag{1.15}$$

where

$$g_{\sigma(\lambda}\hat{R}^{\sigma}_{\mu)\nu\rho} = 0, \quad F_{(\lambda\mu)} = 0,$$

with F being the distant curvature as introduced above.

Proof. A proof is given in Folland (1970), statement E, p. 151; a helpful supplementary resource here is (Nelson, 1967). □

Note that the first term in (1.16) is what Weyl referred to as the 'directional curvature'; the second is what Weyl referred to as 'Streckenkrümmung' (Scholz, 2020, p. 9) (which might be translated as 'distant curvature'—as, again, already introduced earlier—or as 'length curvature').

Lemma. (Vanishing distant curvature criterion for Lorentzian geometry) *In any simply connected domain of M, there is a Lorentzian metric $g_{\mu\nu}$ that is compatible with the conformal structure $\mathcal{C}$ such that the Weyl connection is metric with respect to $g_{\mu\nu}$ iff the distant curvature $F_{\mu\nu} = 0$.*

Proof. Left-to-right: If Γ is metric, then by definition $\nabla_{\mu}g_{\nu\lambda} = 0$, in which case $d\varphi = 0$, and so $F = 0$.

Right-to-left: If $F = 0$, then $d\varphi = 0$, in which case locally $\varphi_{\mu} = d_{\mu}\chi$ for some scalar field χ. But in that case, one is working in an integrable Weyl geometry; and in such a case, one can perform a Weyl gauge transformation (i.e. transformations on $(g_{\mu\nu}, \varphi)$ which fixes the same Weyl geometry)

$$g_{\mu\nu} \mapsto \tilde{g}_{\mu\nu} := e^{f}g_{\mu\nu},$$
$$\varphi \mapsto \tilde{\varphi} := \varphi - f$$

for suitable function f in order to transform to the *Riemann gauge* of the integrable Weyl geometry, in which $\tilde{\varphi} = 0$. One then finds that the metricity condition is satisfied (Scholz, 2020, p. 31). □

The congruence criterion can be established as an immediate consequence (due to its centrality, we still label it as a 'proposition'):

Proposition. (Congruence criterion for Lorentzian geometry) *In any simply connected domain of M, there is a Lorentzian metric $g_{\mu\nu}$ that is compatible with the conformal structure $\mathcal{C}$ such that the Weyl connection Γ is metric with respect to $g_{\mu\nu}$ iff there is no second clock effect.*

Proof. The second clock effect in a Weyl geometry occurs iff there is non-vanishing length curvature $F_{\mu\nu}$. This can be seen by explicitly comparing the result of parallel transporting the same starting vector via two different paths from a point p to a point q: there will only be no change in length iff φ_μ (as defined in the previous proof) is closed, i.e. $d_\mu \varphi_\nu = 0$. But this is again directly equivalent to $F_{\mu\nu} = 0$ (see Bell and Korté (2016)). □

The equidistance criterion comes about by unpacking further what $F_{\mu\nu} = 0$ amounts to:

Proposition. (Equidistance criterion for Lorentzian geometry) *In any simply connected domain of M, there is a Lorentzian metric $g_{\mu\nu}$ that is compatible with the conformal structure $\mathcal{C}$ such that the Weyl connection Γ is metric with respect to $g_{\mu\nu}$ iff equidistant events on a particle P (i.e. geodesics) correspond to equidistant events on a particle P' as described in the equidistance criterion.*

Proof. Right-to-left: Let U be the tangent vector of the first of two adjacent, affinely parameterised geodesics P, P' of a geodesic congruence, and V their connection vector. Consider the corresponding geodesic deviation equation

$$\ddot{V}^\mu = \hat{R}^\mu_{\nu\rho\sigma} U^\nu U^\rho V^\sigma + \frac{1}{2} U^\mu F_{\nu\rho} U^\nu V^\rho. \tag{1.16}$$

with $g_{\sigma(\lambda}\hat{R}^\sigma_{\mu)\nu\rho} = 0$. (For a derivation of the geodesic equation for a torsion-free connection, see Reall (2021), equation (4.22) and its subsequent proof. The specific equation here follows then from splitting up the Riemann tensor as allowed by the above lemma (Curvature decomposition in Weyl structure).

The LHS is $\hat{\nabla}_U \hat{\nabla}_U V^\mu$, which is orthogonal to U^μ since

$$
\begin{aligned}
g_{\mu\nu} U^\nu \left(\hat{\nabla}_U \hat{\nabla}_U V^\mu \right) &= g_{\mu\nu} \hat{\nabla}_U \left(U^\nu \hat{\nabla}_U V^\mu \right) \\
&= g_{\mu\nu} \hat{\nabla}_U \hat{\nabla}_U \left(U^\nu V^\mu \right) \\
&= \hat{\nabla}_U \hat{\nabla}_U \left(g_{\mu\nu} U^\nu V^\mu \right) \\
&= 0.
\end{aligned}
$$

The first term on the RHS is also orthogonal to U^μ: After multiplying with $g_{\mu\zeta} U^\zeta$, one obtains $g_{\mu\zeta} U^\zeta \hat{R}^\mu_{\nu\rho\sigma} U^\nu U^\rho V^\sigma = g_{\mu[\zeta} \hat{R}^\mu_{\nu]\rho\sigma} U^\zeta U^\nu U^\rho V^\sigma = 0$ where the first step uses $g_{\sigma(\lambda} \hat{R}^\sigma_{\mu)\nu\rho} = 0$, and the last step follows from contraction of two antisymmetric with two symmetric indices. At the same time, the second part on the RHS is parallel and not orthogonal to U because it is a scalar multiplied by U (and U is non-null).

Now, if equidistant events on P correspond to equidistant events on P' as described in the equidistance criterion, 'the connection vector V... can be chosen orthogonal to P [i.e. to U.]' (p. 82), because shared equidistance between two curves implies orthogonality. So, choose V to be orthogonal to U. We now hit the whole equation (1.17) with $g_{\mu\nu} U^\nu$: the LHS as well as the first term on the RHS of (1.16) must vanish, as they are orthogonal to U. If this is true for arbitrary U and V, then F has to vanish. From this via the previous lemma it follows that there is a Lorentzian metric $g_{\mu\nu}$ that is compatible with the conformal structure $\mathcal{C}$ such that Γ is metric with respect to $g_{\mu\nu}$.

Left-to-right:[45] Suppose that Γ is metric with respect to $g_{\mu\nu}$. Given this metric compatibility, locally one can find a Fermi normal coordinate system in which connection coefficients vanish along a given geodesic (in that coordinate neighbourhood).[46] Furthermore, in these coordinates, geodesics neighbouring the Fermi coordinate-defining geodesic approximately take the form of straight line trajectories due to continuity, i.e. satisfy $d^2 x^\mu / d\lambda^2 = 0$ for affine parameter λ up to an error (which is a function of how closely the neighbouring geodesic is chosen). Then, the coordinate-defining geodesic (which could have been chosen arbitrarily) and a sufficiently closely chosen geodesic approximately correspond to two straight lines in that coordinate system, from which it follows that Einstein synchrony approximately defines level surfaces of proper time connecting the two geodesics (as familiar from the flat spacetime case), and so the equidistance criterion approximately obtains. Notably,

[45] Note that EPS do not prove the left-to-right direction of the above proof on page 82 of their article.
[46] For background on Fermi normal coordinates, see Manasse and Misner (1963).

the error by which the equidistance criterion is violated is controlled, i.e. it can be made ever smaller the closer the second geodesic is chosen to the Fermi coordinate-defining one. □

Notably, for several statements, the (not further justified) assumption of (local) 'simply connectedness' was required. One such instance, as already pointed out, was in the definition of the notion of distant curvature. The assumption of simply connectedness as such is harmless in so far as it does not exclude any GR spacetimes: it has been shown that any non-simply connected region of an asymptotically flat spacetime must lie behind a horizon (see, for instance, Schleich and Witt (2010)), which then also means that any GR spacetime must be locally simply connected with respect to the manifold topology.[47]

Moreover, no matter which of the above two criteria is used, both can—it seems even must—be operationalised through the use of clocks (after all, in the equidistant criterion, one makes recourse to Einstein conventionality, and, in the congruence criterion, recourse to clocks).

1.7 A preliminary evaluation of the EPS scheme

The EPS scheme sets up the core kinematical structure (M, g_{ab}) of classical Lorentzian spacetime theories—including general relativity—from local statements regarding the propagation of matter and light rays. This construction process of the kinematical structure is additive ('constructivist', in the sense of (Carnap, 1967), as discussed above), so that the resulting kinematical structure (M, g_{ab}) must of necessity be arrived at from the axioms without the possibility of definitional circularity—rather, it is built up linearly therefrom. Insofar as the axioms are justified empirically, the resulting theory will have been placed on firm empirical footing (this, indeed, was also Reichenbach's motivation for the pursuit of the constructive axiomatization of physical theories—see Reichenbach (1969 [1924])). However, insofar as these axioms are not all justified empirically, the empirical status of the resulting structure is not fully guaranteed. As we have seen in our identification of several conventional inputs in the EPS approach, as well as apparently unjustified assumptions and restrictions, this is a more accurate representation of the status of EPS' result.

There are many more points to be made regarding the foundational ramifications of the EPS axiomatization. We close this chapter with a set of comments on (1) limitations and deficiencies, (2) associated amendments, and

[47] Nevertheless, on Hawking, King and McCarthy's path topology, generic spacetimes are not even locally simply connected (see Low (2010)).

(3) overall readings of the EPS scheme. We evaluate constructive approaches to spacetime theories more generally in Chapter 2.

1.7.1 Limitations and deficiencies

Over the course of this chapter, the following limitations and deficiencies of the EPS scheme have become evident:

No dynamics: As it is the purpose of the EPS axiomatization to recover the *kinematical* structure of general relativity, the approach, of course, does not recover the field equations. Presumably, the thought would be that the dynamics can in principle be straightforwardly systematised once the kinematics is in place (cf. Hetzroni and Read (2023)).[48,49]

Locality restrictions: The EPS scheme holds only locally; nothing about global topology of spacetime can be learned via this approach. (Notably, determining the global structure is not even possible from the dynamical sector of GR—so why hold an axiomatization of the kinematical sector accountable then?)[50]

Idealisations: The differentiability assumptions in EPS are an important control handle over possible alternative kinematical settings; they are thus not at all the innocent idealisation assumptions which they appear to be at first sight. Furthermore, the operational assumptions deployed by EPS are highly idealised but—impractical idealisations granted—workable.

Non-constructivist: The EPS scheme arguably deviates from its basic posit of semantic linearity at different points and to different degrees of severity, i.e. strictly speaking the EPS system is thus not a constructivist system in the strict sense. These will be outlined in detail below.

[48] For example, on a (super-)Humean approach, one might begin with a Humean mosaic consisting of the motions of particles and light rays; construct therefrom via the EPS scheme the structure of a Lorentzian spacetime $(M, g_{\mu\nu})$; and then extract the dynamics as the simplest and strongest codification of these (themselves derivative) entities. For discussion of such (super-)Humean approaches, see e.g. (Pooley, 2013a, §6.3.1).

[49] It's in fact not completely obvious that this is the case—for it is not obvious that fixing a Lorentzian metric field *up to a global constant* is sufficient to fix the dynamics. To illustrate: different terms in the Einstein-Hilbert action plus terms associated with different matter fields will feature different orders of coupling to said metric. Scaling by some global constant will, therefore, scale these terms in different ways—thereby not preserving dynamics and moreover not preserving empirical content (for coupling constants will be altered). Our thanks to Brian Pitts for discussion on this point.

[50] For instance, global topology is underdetermined by the field equations. See (Manchak, 2009). In this regard, one might also think that EPS compatibility (*a fortiori* the axioms which secured it) is not completely innocuous, as (for example) there are FLRW spacetimes which are timelike geodesically complete but null incomplete (see (Harada et al., 2018, 2022))—being a local approach, EPS will not obviously enable the construction of such spacetimes. Our thanks to Erik Curiel for drawing this point to our attention. Curiel, indeed, also suggests the following to us: 'delete a point q from Minkowski spacetime, and consider a point p that had the deleted point q in its future null cone; then there is no sequence of future timelike geodesics from p (nor from any other point) that converges to the past-incomplete null geodesics emanating from where q used to be' (p.c.). We agree that the global features of such spacetimes will not be recoverable from EPS alone, *qua* local axiomatic methodology.

Let us consider in more detail the sense in which the EPS axiomatization is non-constructivist. First, the original scheme confirms a common stereotype on how constructivist rational reconstructions are *produced*: if stuck, break your own rule of linearity and simply postulate an *ad hoc* workaround. The bridging criteria from a Weyl metric to a Lorentzian metric offered by EPS—no-second clock effect—seem to take exactly this form: all of a sudden, the availability of clocks is assumed. Notably though, this specific issue need not be conceived as one of circularity: the *ad hoc* move can first of all be seen as a placeholder for a more detailed account on this specific transition. In fact, as to be discussed more in chapter 2, Perlick (1987) has shown how the EPS scheme can be amended at this point by a clock criterion for Weyl space-times that allows for singling out clocks with which one can test for the second clock effect. The proponent of the EPS scheme has simply to concede that the original scheme was incomplete—but need not conclude that it was circular. A similar point can be made in the context of the objection of Sklar (1977) that the EPS approach is circular as it presupposes a distinction between forced versus force-free motion—again, if correct, the results of Coleman and Korte (1980) demonstrate simply that the approach is incomplete, rather than circular. (These additions to the EPS scheme will be assessed in more detail in Chapter 2.)

The original EPS scheme also illustrates a more general stereotype regarding rational reconstructions: one must *fine-tune* the assumptions in the linear derivations in such a way that one gets what one is after, and nothing else. This fine-tuning then brings in a form of *methodological* circularity after all. The differentiability assumptions of EPS, for instance, do not follow from in-principle preconditions on measurement or modelling: they are at most what is pragmatically easiest to work with. Without further justification, they are problematic, as the final result of the rational reconstruction depends heavily on their invocation. Their only justification seems to lie in the fact that we *want* to derive the kinematical structure of general relativity.

Finally, the EPS scheme clearly violates mathematical linearity, i.e. the authors neither carry out—unsurprisingly, given the page constraints of a single paper—nor promote a constructivist mathematical program. It is not clear why linear constructivism when demanded in physics should not be demanded in mathematics as well (although, to be clear, neither Reichenbach nor Carnap advocatre for this). Compare EPS' silence here to the stance taken by the school of 'proto-physics'—composed of Lorenzen, Janich and their students in Germany in the late 1960s to early 1980s—which aims at a non-circular, systematic account of how to set up basic measurement operations for those quantities that are essential to physical theorising—such as distance, time, and mass.[51] Their explanatory standard requiring a linear explanation for measurement operations likewise made them—arguably even as a precondition for their originally-planned undertaking—engage with a linear construction of logic, mathematics, and language more generally.

1.7.2 Amendments

The EPS scheme has been the subject of various variations:

Replacement of non-constructivist axioms As depicted before, non-linear reference to clocks can be remedied through the Weyl structure clock construction à la Perlick (1987). Again, as depicted before, Sklar's non-linearity objection has arguably been addressed by Coleman and Korté (see (Coleman and Korté, 1992), among others).

Alternative basic entities Woodhouse (1973) provides a linear set-up of manifold structure from causality assumptions not so different from that within the EPS scheme; this is an improvement on (parts of) the EPS scheme if any assumption of manifold is seen as to be reduced ultimately to causal relations (the EPS scheme assumes particles to be one-dimensional manifolds; cf. axiom D_1). The scheme of EPS has also been changed to be founded on empirically-motivated axioms regarding alternative probing structures, in particular to particles with spins and matter waves as test particles (see Audretsch and Lämmerzahl (1995) and Audretsch and Lämmerzahl (1991) respectively).

Generalisations through loosening of assumptions Several assumptions (some of them made only implicitly) are not empirically motivated: (i) The implicit assumption of EPS that the manifold satisfies the Hausdorff condition cannot—at least at the point of introduction in the constructional system—be empirically motivated, and may thus be dropped altogether. Upon weakening this assumption, the question becomes whether at some later stage a suitable empirical axiom can be introduced that imposes the Hausdorff condition—or whether the EPS scheme should be widened into a constructive account of the kinematical space of non-Hausdorff GR (see Luc (2020); Luc and Placek (2020)). (ii) In axiom P_2, EPS appear to assume implicitly that coordinate systems are holonomic, thereby excluding torsionful spacetimes by fiat. Without so assuming that coordinate systems are holonomic, one opens the possibility that the axiomatization will yield torsionful spacetimes (although, of course, it is also

[51] Very crudely, its main motivation derives from a dissatisfaction with standard theory-based accounts of measurement, all of which suffer (the claim goes) from circularity in one form or another; its main strategy to evade such circularity lies in grounding measurement semantics in pragmatics. A collection edited by Böhme (1976) (in German only) and a monograph by Janich (2012) serve as good entry points to the literature.

possible that by weakening axiom P_2 in this way, the entire ensuing structure of the axiomatization is modified, thereby yielding further differences in the final class of spacetime kinematical possibilities at which one arrives[52]). (iii) The explicit assumption that g, as defined in the axiom, is a function of class C^2 on U is (again) a strong idealization. Upon weakening the assumption, the EPS scheme can be deformed into a constructive scheme for at least a subclass of Finslerian spacetimes (of which Lorentzian spacetimes are a special case): see Pfeifer (2019), section II C, and also Lämmerzahl and Perlick (2018).

1.7.3 Readings

In light of the above walkthrough to the EPS construction, we note the following specific readings (note that these are options which are intended to be neither mutually exhaustive not mutually exclusive):

EPS as a what-is-sufficient explanation of kinematical structure: According to Ehlers (1973, pp. 22–23),

> [the] main purpose [of the EPS scheme] is to elucidate from a different point of view and in a systematic manner, which facts and mathematical idealizations of facts are sufficient to support the edifice of general relativity theory. In particular the following approach enables one to isolate certain relatively independent substructures which are contained in Einstein's theory, the differential-topological, conformal, projective, affine and metric structures, which are related to different observable phenomena or to different relations inferred from such phenomena. These substructures are present also in theories different from Einstein's and might also be considered as building blocks for further developments needed when Einstein's theory has to be modified.

This arguably complements a what-empirical-structure-is-necessary explanation of kinematical structure through the usual theoretical account which implies all of the used axioms by EPS.

EPS as a how-possible explanation of theory change: Although it is not clear from the EPS scheme that the probing assumptions are meant necessarily as probing assumptions from a local special relativistic spacetime,[53] one might, in any

[52] Notably, the assumption of torsion-free affine connection is also used in showing the equivalence between constructive and algebraic notion of projective structure in the set-up; dropping it might thus also be problematic at this point.

[53] At least, light is tacitly assumed to propagate with a finite propagation speed in the EPS scheme.

case, argue that the EPS scheme can be understood to provide a conceptualisation of how theory change is possible from one theory to another by showing how the localisation on probing assumptions leads to novel kinematical structure (Lorentzian spacetime structure instead of Minkowski spacetime). Thus read, the EPS scheme can be seen to provide a *how-possible* explanation of how a shift from special to general relativistic kinematics could have come about—how, in other words, one can 'bootstrap' from a given initial set of theoretical commitments to some other, richer, set of theoretical commitments.[54]

EPS as an ontological account of spacetime structure The EPS programme (at least when freed of circularity) can, potentially, be re-construed as ontologically reducing spacetime structure to an ontology of free particles and light rays. That said, strictly speaking there are problems for this reading: what we in fact seem to require in order for this ontological reduction to succeed is a reduction to dense sets of *possible* free particle trajectories and light rays, which would seem to introduce a modal dimension which may be repugnant to the empiricist agenda underlying the programme of constructive axiomatics. (The approach—construed ontologically—would thereby seem to qualify as a version of 'modal relationalism', on which see (Belot, 2011); we will discuss this issue further in Chapter 3. Note also that the approach of using certain pieces of basic ontology to fix further ontological commitments is in the spirit of the Lewsian approach to functionalism recently defended in the philosophy of physics by Butterfield and Gomes (2023).)

In any case, setting these concerns aside: albeit *prima facie* just one out of several options to provide a fundamental ontology to GR, the ontological basis suggested by the EPS scheme may be singled out by a view of naturalised metaphysics according to which a fundamental ontology primarily is to be simple and informative.[55] And of course, it bears stressing that this ontological focus is distinct from the often-declared *epistemological* focus of proponents of constructive advocates ('given such-and-such empirically unproblematic starting points, to what extent can I access—and thereby render unproblematic empirically—the edifice of my theory in general?')

Furthermore, there is the more intricate question of how to relate the EPS scheme to past programs of linearly constructing spacetime structure out of more primitive structure, such as (a) Klein's Erlanger program, (b) Reichenbach's constructive axiomatization, (c) protophysics, and (d) Brown's dynamical approach and other accounts of physical geometry. These will be some of the major topics of Chapter 2.

[54] Cf. the how-possible explanations of how general relativic kinematics can be motivated from special relativistic kinematics in (Mittelstaedt, 2013, Ch. 2); cf. also the discussion of the heuristics of the move from special to general relativity presented recently by Hetzroni and Read (2023).

[55] Cf. Esfeld (2020) and other proponents of the minimalist ontology approach to quantum mechanics and physics more generally. See also our remarks on (super-)Humeanism in footnote 48.

1.A Summary of the EPS axioms

In this appendix, for ease of reference, we collect together in one place all of the EPS axioms. Some of these axioms deploy defined terminology, for which the reader is referred to the body of this chapter.

Differential topology axioms

Axiom (D_1). *Every particle is a smooth, one-dimensional manifold; for any pair P, Q of particles, any echo on P from Q is smooth and smoothly invertible.*

Axiom (D_2). *Any message from a particle P to another particle Q is smooth.*

Axiom (D_3). *There exists a collection of triplets* (U, P, P') *where* $U \subset M$ *and* $P, P' \in \mathcal{P}$ *such that the system of maps* $\{x_{PP'}|_U\}$ *is a smooth atlas for M. Every other map* $x_{QQ'}$ *is smoothly related to the local coordinate systems of that atlas.*

Axiom (D_4). *Every light ray is a smooth curve in M. If* $m : p \rightarrow q$ *is a message from P to Q, then the initial direction of L at p depends smoothly on p along P.*

Light axioms

Axiom (L_1). *Any event e has a neighbourhood V such that each event p in V can be connected within V to a particle P by at most two light rays. Moreover, given such a neighbourhood and a particle P through e, there is another neighbourhood* $U \subset V$ *such that any event p in U can, in fact, be connected with P within V by precisely two light rays* L_1, L_2*, and these intersect P in two distinct events* e_1, e_2 *if* $p \notin P$*. If t is a coordinate function on* $P \cap V$ *with* $t(e) = 0$*, then* $g(e) := -t(e_1)t(e_2)$ *is a function of class* C^2 *on U.*

Axiom (L_2). *The set* L_e *of light-directions at an (arbitrary) event e separates* $D_e \setminus L_e$ *into two connected components. In* M_e *the set of all non-vanishing vectors that are tangent to light rays consists of two connected components.*

Projective axioms

Axiom (P_1). *Given an event e and a* $\mathcal{C}$*-timelike direction D at e, there exists one and only one particle P passing through e with direction D.*

Axiom (P_2). *For each event* $e \in M$*, there exists a coordinate system* $(\overline{x}^{\mu})$*, defined in a neighbourhood of e and permitted by the differential structure introduced in axiom* D_3*, such that any particle P through e has a parameter representation* $\overline{x}^{\mu}(\overline{u})$ *with*

$$\left.\frac{d^2\overline{x}^{\mu}}{d\overline{u}^2}\right|_e = 0;$$

such a coordinate system is said to be projective at e.

Compatibility axiom

Axiom (C) *Each event e has a neighbourhood U such that an event* $p \in U$*,* $p \neq e$ *lies on a particle P through e if and only if p is contained in the interior of the light cone* v_e *of e.*

2
Constructive Axiomatics in Context

Introduction

In his 1924 book *Axiomatization of the Theory of Relativity*, Hans Reichenbach proposed the method of constructive axiomatics upon which EPS would subsequently build. Recall from the previous chapter that according to this, physical theories, such as Einstein's theory of relativity, should be formulated in such a way that their content is built up from axioms which have direct and indubitable empirical content; in this way, the physical content of the theory under consideration is (supposedly) rendered manifest (Reichenbach 1969 [1924]). Here is Reichenbach himself on the idea:

> It is possible to start with the observable facts and to end with the abstract conceptualization. A certain loss in formal elegance will be balanced by logical clarity. The empirical character of the axioms is immediately evident, and it is easy to see what consequences follow from their respective confirmations and disconfirmations. Such a *constructive* axiomatization is more in line with physics than a *deductive* one [that runs from the abstract to the observable], because it serves to carry out the primary aim of physics, the description of the physical world. (Reichenbach (1969 [1924], p. 5))

The idea of constructive axiomatics has since found its way into the broader literature on the foundations of physics. Recall again the following characterization from Carrier on constructive axiomatics, already presented in the Introduction:

> In a constructive axiomatization only those statements are accepted as fundamental that are immediately amenable to experimental control. A deductive axiomatization, by contrast, confers fundamental status to more abstract statements (such as variational principles). (Carrier (1990, p. 370))

It is important to clarify at this early point that a constructive axiomatization need not be formulated explicitly through terms which refer to what is "immediately amenable to experimental control" (call such terms 'constructive').

Constructive Axiomatics for Spacetime Physics. Emily Adlam, Niels Linnemann, and James Read, Oxford University Press. DOI: 10.1093/9780198922391.003.0003

Rather, also any term can be used that is defined linearly through constructive terms and/or has already been linearly defined through constructive terms. So, to stress: constructive axiomatization employs constructive terms and terms in line with what one could call a postulate of semantic linearity (Carrier, 1990). We call 'constructivist' any axiomatization that follows a posit of semantic linearity—i.e., for which the permissible vocabulary consists of a base vocabulary plus all terms that can be defined linearly through the terms of the base vocabulary and/or terms that have already been defined linearly through terms of the base vocabulary—independently of whether the basic vocabulary used is constructive or not. Importantly, the noun 'constructivism' will only be used with respect to this adjective 'constructivist'.

Now, Reichenbach's attempt to axiomatize the theory of relativity was not wholly successful. In the context of the special theory of relativity, for instance, he failed to appreciate that the paths of light rays—in terms of which he expressed his constructive axioms—do not in themselves suffice to recover the theory: one needs, in addition, either (i) projective structure, which is associated with the paths of massive particles, or (ii) certain global topological assumptions (regarding the compactification of $\mathbb{R}^4$ by a light cone at conformal infinity—for the details and history on this point, see Rynasiewicz (2005)). Such issues were pointed out repeatedly to Reichenbach by Weyl (who would go on to present his own constructive approach to relativity theory), but were—much to Weyl's chagrin—apparently ignored.[1]

In the context of the general theory, Reichenbach sought to use idealized rods and clocks in local neighbourhoods to recover the metric field: but one might reasonably complain that (i) his use of 'local neighbourhoods' is not sufficiently mathematically well-defined,[2] and (ii) in a constructive

[1] Again, see Rynasiewicz (2005) for a fascinating history of this episode. Curiously, in the modern context Knox (2013) seems to make a similar mistake to Reichenbach, but with respect to particle trajectories rather than light rays:

> in a relativistic context [. . .] if we know the full set of geodesics (the inertial trajectories) then this is sufficient to fix both conformal and projective structure. This in turn fixes metric structure up to a global scale factor (not just a conformal factor), and thus determines all relative distances, which I take it is sufficient to fix the operational content of the metric. (Knox, 2013, p. 349)

It is important to note that a full set of geodesics does *not* suffice to fix a Lorentzian metric structure; for this—per the EPS existence result—one also needs the trajectories of light rays, plus various further inputs. What Knox presumably has in mind is that the 'timelike' nature of geodesics suffices to demarcate null cone structure at every point and thus ultimately indeed conformal structure (just think of the null cone at point p as defined as boundary to the set of timelike geodesics going through that very same point). Although this idea is legitimate, it evidently goes beyond anything which EPS have in mind (*pace* Knox, who also seems to impute the claims in the above passage to EPS).

[2] Such concerns persist a century later—see e.g. Weatherall (2021) in response to Read et al. (2018).

approach, one cannot *presuppose* the existence of such complex bodies as rods and clocks.[3] Again, both of these issues were (seemingly) overcome by Weyl, who was able to show that a given Lorentzian metric field is fixed (up to a spacetime-dependent scale factor) by its projective structure (associated with the paths of massive bodies) and conformal structure (associated with the paths of massless bodies) without any need to make recourse to rods or clocks (Weyl, 1921). However, it would in fact be fifty years until a constructive axiomatization of general relativity in the spirit of Weyl was articulated fully.

The project of the constructive axiomatization of general relativity can be seen as being taken up once again in 1972 by Ehlers, Pirani, and Schild, in an attempt to counter the by-then dominant *deductive* chronometric method. On the chronometric method, as explicitly first and foremost championed by Synge (1959; 1960), good clocks are *stipulated* to read off the worldline interval of the path along which they travel.[4] The approach is deductive, as testable hypotheses—ultimately statements on the worldline intervals of a path—are derived from the *readily presupposed* theory of general relativity.

In contrast to this methodology, EPS took up Weyl's above-mentioned result in an attempt to complete the constructive project. Prima facie, no clock is needed on this approach to establish the metric structure of spacetime, because Weyl's theorem is understood to provide a direct way to link much of the theoretical structure of general relativity to local experience (namely via the allegedly relatively theory-neutral notions of free particles and light rays) without recourse to theory-external proxies such as clocks, as they are evidently used in the chronometric method.[5]

Building on Weyl's work, EPS subsequently managed to show that the causal-inertial method can even serve as a suitable basis for a constructive axiomatization: all relevant structure from the differentiable manifold up to the metric was more or less rigorously shown to follow step-by-step from

[3] Famously, this was one of the later Einstein's later misgivings about the theory of special relativity as presented in his 1905 paper (Einstein, 1905)—for the history here, see (Giovanelli, 2014). In 1919, Einstein identified his 1905 formulation of special relativity as a 'principle theory'—which is a theory built from observed empirical regularities, raised to the status of postulates (Einstein, 1919). There are clear affinities between principle theories and constructive axiomatizations—discussed in detail later in this chapter—so it is of little surprise that the 'cannot presuppose complex bodies' complaint should arise in both cases.

[4] In modern-day parlance, this assumption is tantamount to what is known as the *clock hypothesis*: see Maudlin (2012, p. 76) for discussion.

[5] This point was made by Weyl himself: see Weyl (1922, p. 63), and so not at all novel to the constructive approaches yet to come. In spite of saying this, it is important not to confuse EPS' achievement with Weyl's theorem—a point which we stress several times throughout this book.

plausible starting assumptions and observation on the local behaviour of light and free particles; as a substantial part of this, the EPS approach establishes the *existence* of a Weyl structure given suitable projective and conformal structures. Compare this to Weyl's original work: the theorem by Weyl—usually referred to as 'Weyl's theorem' in the current context—states only that the Weyl structure is uniquely *determined* by the projective and conformal structure of a presupposed metric; so, Weyl's theorem is a uniqueness result about a Weyl metric, but is not an existence result—let alone an explicit construction of a Weyl metric.[6]

As indicated already in Chapter 1, over the years the EPS scheme has found amendments in various directions. Most importantly, proposals for a more admissible basis for GR's constructive axiomatization have been put forward (for instance, by (Woodhouse (1973)) and (Audretsch and Lämmerzahl (1991))), and problematic steps in the original scheme remedied. With respect to the latter point, in particular, the following achievements are worth highlighting: (1) an arguably highly problematic circularity in determining projective structure was (supposedly) addressed by (Coleman and Korté (1992)); and (2) the apparent reliance of the EPS scheme on primitive clocks in determining the conformal factor was (supposedly) circumvented through a clock criterion in Weylian (affine) space by (Perlick (1987)) (cf. (Schmidt (1995))). An important (albeit somewhat irritating) result raised in discussions of the EPS scheme over the years consists furthermore in the fact that (3) the details of the constructed spacetime structure depend crucially on certain geometrical (arguably idealizing) assumptions which are hard to justify. For instance, for a given set of local observations, the EPS scheme can lead to some Finslerian-non-Lorentzian spacetime if the twice differentiability assumption of the echo function $p \to t_e \cdot t_r$ (t_e and t_r are emission and return 'times' respectively) is dropped (see (Pfeifer (2019); Lämmerzahl and Perlick (2018))).[7]

Our purpose in this chapter is to provide a comprehensive appraisal of the potential—but also the limitations—of the constructive axiomatic methodology as exhibited by EPS. For this, we take stock of the plethora of variants on the causal-inertial methods, and related constructive approaches to spacetime theories. We focus on constructive approaches to general relativity, but some of the results pertain to neighbouring or generalized theories of general relativity

[6] This existence result follows rigorously only when one supplements EPS with recent results from Matveev and Scholz (2020).

[7] This issue was also raised in Chapter 1.

(e.g. teleparallel gravity) just as well. In fact, our undertaking can be read as a case study on the merits of a constructive approach to a physical theory more generally.

The plan for the chapter is this. §2.1 discusses the 'causal-inertial' constructive axiomatic take on general relativity, with (as discussed above) its roots in Weyl's theorem, and taking its mature form in the EPS axiomatization. After clarifying the relationships between Weyl's work and the EPS approach, we introduce and assess the (supposedly) circularity-free refinement of EPS developed by Coleman and Korté. In §2.2, we consider the interplay between chronometry (i.e. appeals to clocks) and constructive axiomatics: is it indeed the case that constructive axiomatics such as those of EPS need make no appeal to such complex constructions, or is their use, rather, implicit in the approaches—and if so, where is it implicit? §2.3 discusses various notions of constructive axiomatization and constructivism. We consider (i) the extent to which constructivist approaches may legitimately make reference to theoretical notions associated with external theories (i.e. drop the constructive but keep the constructivist nature), and (ii) whether there could be a sense in which constructive axiomatization can proceed via more general iterative, rather than just linear, methodologies (i.e. drop the constructivist but keep the constructive nature). The section also treats the question of (iii) how constructive axiomatics relates to the idea of the 'Einstein–Feigl completeness' of a theory, which is roughly the idea that a theory contains its own observation theory: an issue previously discussed by Carrier (1990). §2.4 situates constructive axiomatics in the broader philosophy of spacetime and of physics, by drawing connections between constructive axiomatics and a number of other programmes and issues. §2.5 raises in the context of constructive axiomatics the general epistemological question of how viable foundationalism about a physical theory can be *per se*.

2.1 Causal-inertial-type constructivism

In this section, we present and discuss three important projects which form part of the history of the EPS scheme. In §2.1.1, we explain the exact sense in which this approach is related to a series of theorems by Weyl, proved in the 1920s. In §2.1.2, we consider the attempt by Coleman and Korté to repair an apparent circularity in the EPS approach, and consider whether these authors' revised version of EPS evades that circularity problem. In §2.1.3, we discuss Perlick's proposed amendments to the EPS program.

2.1.1 The original motivation: Weyl's theorem

The constructive axiomatic route of EPS is in its motivation heavily indebted to a result by Weyl, according to which the Weyl metric on a differentiable manifold is uniquely determined by its projective and conformal structure:[8]

> Projective and conformal structure of a [Weyl] metric space uniquely determine the [Weyl] metric (Weyl (1921), Satz 1)[9,10]

The result guarantees that if suitable projective structure and conformal structure that give rise together to a Weyl metric are found, *then* the Weyl metric will be determined uniquely by these. Notably, the result is not an existence result for a Weyl metric in the presence of conformal and projective structure.

Here is a sketch of a proof of Weyl's theorem (to repeat, proved in (Weyl, 1921; Weyl and Delphenich, trans.); cf. (Scholz (2020)). Consider a Weyl manifold $(M, [(g_{\mu\nu}, \varphi_\mu)])$, with $g_{\mu\nu} \mapsto \tilde{g}_{\mu\nu} = \Omega^2 g_{\mu\nu}$ and $\varphi_\mu \mapsto \varphi_\mu - d_\mu \ln\Omega$ as usual (and as defined in Chapter 1). The difference tensor between any two derivative operators compatible with the metrics in this class is given by

$$\Delta^\mu_{\nu\sigma} = -\varphi_{(\nu}\delta^\mu_{\sigma)} + \frac{1}{2}\varphi^\mu g_{\nu\sigma}, \tag{2.1}$$

which is a result by now familiar from the previous chapter (cf. also (Wolf et al. (2023a, §4.2)). Now, projective structure fixes the first term in the above equation, while conformal structure fixes the second; together, this means that the connection of the Weyl manifold and—since the connection fixes the Weyl metric up to global constant scaling—the Weyl manifold as such is indeed fixed by its associated projective and conformal structures.

[8] Kretschmann (1917) worked on this as well, and might even have arrived at a similar result prior to Weyl (see Coleman and Schmidt (1995), p. 1343).

[9] The authors' translation of the German original: 'Projektive und konforme Beschaffenheit eines [Weyl-]metrischen Raums bestimmen dessen [Weyl-]Metrik eindeutig' (Weyl (1921), Satz 1). Although other propositions (*Sätze*) are proven in this paper, it is Satz 1 which generally is referred to as 'Weyl's theorem' in this context.

[10] As an immediate corollary it follows for Lorentzian metrics:

Corollary. *Assume $g'_{ab} = \Omega^2 g_{ab}$. Further, assume g'_{ab} and g_{ab} agree as to which smooth, timelike curves can be reparameterized so as to be geodesics. Then Ω is constant. (Malament, 2012, prop. 2.1.4, p. 127)*

(Recall again n. 7 regarding the distinction between unparameterized geodesics and parameterized geodesics.) Note that g'_{ab} and g_{ab} agreeing as to which smooth, timelike curves can be reparameterized so as to be geodesics is equivalent to their agreeing on unparameterized geodesics, i.e. is equivalent to their being associated with the same projective structure.

Now, the causal-inertial method—as the name already suggests—aims to construct from causal structure (in the sense of conformal structure) and inertial structure (in the sense of projective structure) a (Weyl) metric structure. In the EPS approach, just as the Weyl metric is seen as determined uniquely by its projective and conformal structure (Weyl's theorem), projective and conformal structure are seen as determined by their respective geodesic structure. It is important to stress the distinction of the latter existence result from the former uniqueness result (which is behind Weyl's theorem); the existence result is a core result of the EPS scheme: provided that there is projective and conformal structure such that the null lines of the conformal structure are geodesics of the projective structure, both structures together define a Weyl structure.[11]

Now, in relation to EPS, it is interesting to note that conformal structure in particular is determined uniquely by (at least) one other criterion than the corresponding structure of null geodesic paths—namely, by the path structure of all timelike curves (Malament, 1977).[12] How do the two options compare operationally? (Let us assume that the only operationally sensible method to measure out events from one specific location is via radar coordinates, and that the only reliable standard of constant signal speed in curved spacetime is that of light.) One advantage of this new 'timelike curve' option over EPS' original approach lies in the fact that it is operationally easier to measure out the paths of timelike curves in such radar coordinates than the paths of light-like signals which themselves move at the speed of light. Disadvantages of this new approach, on the other hand, seem to dominate: firstly, it might mean (but this has to be studied with care) that many more curves have to be probed to obtain a good resolution of the conformal structure (figuratively speaking, the

[11] To repeat, the existence result has been been given a rigorous formulation by Matveev and Scholz (2020).

[12] In total, and more formally, two metric fields g'_{ab} and g_{ab} are conformally equivalent iff

Light cone structure g'_{ab} and g_{ab} agree on which vectors, at arbitrary points of M, are timelike (or agree on which are null, or which are causal, or which are spacelike).

Timelike path structure g'_{ab} and g_{ab} agree on which smooth curves on M are timelike. (This is Malament's famous 1977 result, see Malament (1977), which improves on a previous result by Hawking et al. (1976)).

Null geodesic structure g'_{ab} and g_{ab} agree on which smooth curves on M can be reparamaterized so as to be null geodesics.

(This is a combination of the proposition 2.1.1. (p. 122), and proposition 2.1.2. (p. 125) from Malament (2012)).

inside of the light cone has to be measured out as opposed to just its boundary); secondly, and arguably more pertinently, the actual universe contains much more light (think of the CMB for instance) than actual masses: working with actually available light rays is thus a live option for determining conformal structure.[13]

2.1.2 Circularity-free EPS: Coleman and Korté's account

In the wake of EPS' paper, the charge was quickly raised that the approach suffers from a fatal circularity, insofar as it requires substantial conventionalist *inputs* regarding spacetime geometry. The most immediate charge of conventionalism—due originally to Sklar (1977)—is with respect to the status of inertial structure:

> Is a particle free or not when it is gravitationally attracted by another particle? If we say it is, and use the EPS construction, we shall get one spacetime. But what if we say it is not? Won't we get a whole 'conventionally alternative' spacetime by using the EPS construction? (Sklar (1977, p. 259))

Essentially, the issue is this: how is one to identify the *freely-falling* particles—an input in the EPS construction—without an antecedent understanding of spatiotemporal structure used to underwrite this notion? Motivated by this concern, Coleman and Korté developed, over the course of a series of papers, a (supposedly) non-circular modification of the EPS axiomatization. In this section, we provide a brief summary of Coleman–Korté's 'improvement' of EPS, as presented in particular in (Coleman and Korté, 1980) and (Coleman and Korté, 1992),[14] as well an appraisal thereof.

Letting the acceleration of a particular point particle (in the terminology of Coleman and Korté, 'monopole') be $\xi^b \nabla_b \xi^a$, where ξ^a is the velocity vector of the particle under consideration, and ∇ is a derivative operator on the differentiable manifold, Coleman and Korté introduce a *directing field* Ξ^a, as that term on the right-hand side of Newton's second law:

$$\xi^b \nabla_b \xi^a = \Xi^a. \tag{2.2}$$

[13] We thank an attentive reviewer for this suggestion.
[14] See also the excellent summary by Coleman and Schmidt (1995).

How, though, is the directing field to be identified operationally? In order to answer this question, Coleman and Korté (1995b, p. 176) introduce what they dub the 'monopole criterion':

> If two monopoles of the same class are launched at nearly the same event with nearly the same 3-velocity, then under all external physical conditions, their (future) worldline paths will be nearly the same.

By 'monopoles of the same class' are meant monopoles subject to the same directing field on the right-hand side of (2.2). The monopole criterion, then, states that if two particles (monopoles) are assigned the same directing field, then they must follow approximately the same trajectories, for approximately the same initial conditions. Even more simply: particles subject to the same forces and with roughly the same initial conditions should behave in roughly the same way.[15] Coleman and Korté (1995b, 3.19–3.20) present a procedure via which, given the monopole criterion, directing fields can be constructed; and, given what they call the 'cubic criterion', specific directing fields can be singled out as 'geodesic'. The following case distinction then applies:

> If extensive investigation failed to reveal even a single class of particles governed by a geodesic directing field, then the EPS construction would fail to demonstrate the existence of a unique [pseudo-]Riemannian metric. Such a structure might still exist, but other means would have to be sought to establish evidence for its existence. If one or more classes of particles governed by geodesic directing fields were found and if none of the projective structures were compatible with the conformal structure, the construction would fail as before. If two or more projective structures were found which were compatible with the conformal structure, then not even a unique Weyl structure would exist let alone a unique [pseudo-]Riemannian structure. There remains the case in which exactly one class of particles governed by a geodesic directing field compatible with the conformal structure is found. Then the projective path structure revealed by these particles and the conformal structure revealed by light propagation together determine a unique Weyl structure. (Coleman and Korté, 1980, p. 1351)

[15] Note that, at least in Coleman and Korté (1995a,b), Coleman and Korté do not assert the reverse implication: that particles which behave in the same way should be subject to the same forces. This will be of relevance below.

So far so good. To what extent then does Coleman and Korté's construction satisfactorily address Sklar's objection? Now, the methodology allows for finding a unique inertial structure, *provided that there is one to begin with*; no more could be asked from an epistemological approach.[16] Thus, clearly, EPS cannot be counted as epistemologically circular the way Sklar seems to suggest.

It is important to stress that the methodology, albeit fallible, is such that if it fails it does so in an immediately noticeable fashion. This allows us to contextualise the following complaint by Pitts:

> Unfortunately the anti-conventionalist argument hinges entirely on the assumption that there is exactly one physically relevant conformal metric density and exactly one physically relevant projective connection. (Pitts (2016, p. 84))

There are two ways of reading this: either Pitts claims that the Coleman–Korté anti-conventionalist methodology as such presupposes the assumption that there is a unique standard of projective structure—but then his point is just wrong: no such assumption is made; it is just that (and noticeably so) the methodology proposed by Coleman and Korté breaks down if there is not a unique standard of projective structure. Or he indeed simply wants to hint at the limited scope of the Coleman–Korté methodology—which is, however, not an epistemological circularity but simply an epistemological limitation.

Now, Coleman and Korté assert that monopoles in the same class and with approximately the same initial conditions should behave in approximately the same manner. One might thus rather pettily remark that they do not, however, assert the converse, i.e. that monopoles with approximately the same initial conditions which behave in approximately the same manner must be assigned to the same monopole classes—i.e. they do not assert that such monopoles must be regarded as being subject to the same directing field. In this case, again, there arises conventional choice as to whether to regard monopoles as belonging to the same class, and so there arises conventional choice about the fields and objects which, ultimately, correspond to inertial structure. Now, unlike Coleman and Korté (1995a, b), at (Coleman and Korte (1980, p. 1350)) (an older piece) this condition *is* asserted as a biconditional—in which case, the second of these two concerns does not apply. A charitable reading here would

[16] We should also flag that there is a range of less mathematically heavy-duty approaches to identifying inertial trajectories (i.e. to identifying the freely-falling bodies): see Read (2023a, ch. 1) for a recent review.

be to take Coleman and Korté to be deploying the 'mathematician's if', that is, to adhere to the practice of mathematicians to literally formulate definitions as 'if statements' even though they are strictly meant as biconditionals. Even if this were not what the authors had intended to say here, it does strike us that the assumption of the converse direction is indeed natural for parsimony reasons (after all, why have more monopole classes than observationally distinguishable?).

There are, however, two kinds of conventionalism that operate at a higher level and which cannot be blocked so easily by something like the Coleman–Korté approach. First, there is an (apparent) programmatic circularity in setting up selection criteria for geodesics (as proposed by Coleman–Korté), identified by Carrier:

> The theory in fact enters the scheme even if only in a hidden or indirect fashion. Why, after all, do we regard trajectories influenced by gravitational quadrupole moments as distorted whereas we consider paths determined by the action of gravitational monopoles as perturbation-free? Evidently, it is the explanatory theory (GTR) that induces and justifies this judgment. Theory enters constructive axiomatization by *defining the ideal cases.* It supplies *criteria of adequacy* for our choice of basic processes. Only because we already know GTR we can feign to be ignorant of it. (Carrier, 1990, p. 380)

Carrier's point is that one only adheres to a correction scheme such as that of Coleman–Korté if one already has a reason for preferring certain idealized free-fall structure—as opposed to some structure distorted relative to it—as decisive.[17,18] To stress: Carrier's complaint is not that, once an improvement scheme *à la* Coleman-Korté has been chosen, it is itself in its application circular; rather, the circularity which he identifies exists (supposedly) at the programmatic level. Now, in fact, it is not at all clear that a programmatic dependence on some desired target theory—as is arguably the case here—truly

[17] Carrier (1990, pp. 379–80) distinguishes two kinds of correction: by a theory different from that considered specifically (e.g. electrodynamics), and by the theory considered itself.

[18] Cf. Hilbert's thinking on the purpose of axiomatizing physical theories, summarised by Corry as follows:

> The edifice of science is not raised like a dwelling, in which the foundations are first firmly laid and only then one proceeds to construct and to enlarge the rooms. Science prefers to secure as soon as possible comfortable spaces to wander around and only subsequently, when signs appear here and there that the loose foundations are not able to sustain the expansion of the rooms, it sets about supporting and fortifying them. This is not a weakness, but rather the right and healthy path of development. (Corry (2004, p. 127))

For a further fascinating discussion of Hilbert on these issues, see Corry (2018).

deserves the label 'circularity' (of whatever form): after all, it is not the case that a constructive axiomatic endeavour leads to a specific (family of) theories just because *some* structure is construed with *some* goal theory in mind at *some* intermediate step of the overall theory construction. In other words, the programmatic circularity attested to the improved EPS scheme seems, in our view, to be non-vicious (albeit not fully in the desired constructivist spirit).

Second, a global conventionalism which concerns the complete reformulability of theories into empirically equivalent alternative theories (with a different formal set-up to begin with) is not evaded by EPS or in a constructivist project of its kind: the constructivist has to start with some basic way of formalizing the world (in the EPS case, in terms of basic differential-topological statements) which is possibly a conventionalist choice—possibly there are empirically equivalent formalisms available to capture the basic empirical insights without differential-topological structure (see e.g. the alternatives to using standard manifold differential-topological structure discussed in (Fletcher, 2021, §3.1)).

2.1.3 The local constraints of the radar method

The radar method is in general only available in a neighbourhood around a point. How practical, then, is the radar method? Before we attempt to answer this question, note that there are basically two distinct radar methods referred to in the context of EPS: (i) that of EPS, and (ii) that in the spirit of Perlick (1987). While that of EPS requires two observers, that of Perlick is a straightforward one-observer procedure: radar method (i) fixes points through the intersection of emitted and received signals of one observer with that of another, whereas radar method (ii) fixes points through the intersection of an emitted signal (that is, implicitly, taken to be singled out further through two directional angles) with the reflected signal from those points alone.

One might simply claim that Perlick's scheme is thereby more tractable operationally than the original EPS radar method: the problem with method (i), after all, is that it requires communication of the two-observer systems: each observer must send (with some appropriate encoding) to the other observer their two measurement values (i.e. emission and reception times) so that the other observer can complete their respective set of coordinates for the event in question (which consists of the respective emission and reception times on both observers' worldlines for the event in question). Notably, though, an intra-observer communication becomes necessary even on Perlick's account of radar coordinates (method (ii)) as soon as the EPS scheme is meant to reconstruct more than just the local kinematical arena of one observer.

How, for instance, to relate two radar coordinate maps of the Perlick type associated with a pair of neighbouring, arbitrarily parameterized wordlines in some generic spacetime? A straightforward procedure might run as follows:

1. Send a light ray L_1^E from worldline γ_1 towards wordline γ_2; let the light ray hereby encode the emission parameter time t_1^E.[19]
2. Record the parameter time of reception for L_1^E on γ_2 (denoted by t_2^R).
3. Immediately send back a light ray from γ_2 as the reflection of L_1^E (denoted by L_1^R); let this light ray L_1^R hereby encode the values of t_1^E as well as of t_2^R.
4. Record the parameter time of reception at γ_1 (denoted by t_1^R).
5. The observer at γ_1 can now associate an event on γ_2 with two of her own parameter values for emission and receival (t_1^E, t_1^R) as well as the parameter value of that event for γ_2 (t_2^R).

Especially noticeable in significantly non-static spacetimes—i.e., spacetimes which are not approximately static on the relevant timescales—is the fact that one would have to constantly keep tiling spacetime with such a light-signal exchange in order to guarantee the interchangeability between neighbouring radar coordinates, which is, admittedly, far from practical. Again, it bears stressing that the encoding proposed here represents yet another quite explicit instance of *external* theory-dependence.[20]

2.2 Chronometry and constructivism?

Chronometric approaches to GR treat clocks as primitive objects, whereas the proponents of constructive axiomatizations such as EPS typically claim that their methodology does not require us to take objects of this kind for granted. In their introduction, Ehlers et al. (2012 [1972]) give three reasons as to why they deem the chronometric approach unsatisfactory for a *constructivist* project—in particular vis-à-vis one via free particles and light rays:

[19] Some external theory will describe the encoding—so theory-neutrality is also lost here. Note that this is at least in line with Reichenbach's most permissive conception of constructive axiomatics according to which the constructive vocabulary may be subject to theoretical terms as long as the employed terms are different from the theory to be constructed (Reichenbach (1969 [1924], pp. 6–7)).

[20] As we have already seen in the above note, though, in a constructivist account *à la* Reichenbach (1969) this is unproblematic.

(1) Clock readings alone do not seem sufficient for motivating the Lorentzian metric line element $ds = \sqrt{-g_{\mu\nu}dX^{\mu}dX^{\nu}}$ at each point—rather than, say, some first-order line-element $ds = g_{\mu}dX^{\mu}$. For instance, they claim that recourse to light structure (arguably to capture the light cone structure so characteristic of the Lorentzian line metric element) would have to complement the clock readings for motivating a Lorentzian line element. (2) Relatedly, clock readings do not seem to establish geodesic structure in general, that is neither for light nor for massive particles. (3) At the same time, clocks can be constructed in terms of the geodesics of massive particles and light rays (as suggested for instance by the construction of Marzke and Wheeler (1964)[21]), prompting the claim that the geodesics of massive particles and light rays 'alone already imply an interpretation of the metric in terms of time' (p. 65). (According to EPS, brute associations with external clock readings are only needed to link a notion of time to a notion of time of certain matter ('atomic clocks').) We would like to add that, in the spirit of Einstein and Feigl (as we will also elaborate on in the following), one might also take issue with a brute association of the structure of the theory (the worldline interval) with the readings of a theory-external (complex) device that is then not further analysable in the terms of the actual theory at stake.

However, it has been argued that EPS and other constructive axiomatizations do in fact appeal implicitly to clocks—in particular, in the course of moving from a Weyl metric to a Lorentzian metric EPS using a 'no second-clock' criterion which arguably presupposes the availability of clocks as primitive objects. In this section, we consider various approaches to making sense of clocks (and also rods) in a constructive axiomatization, and assess the extent to which they can be employed within the EPS scheme in order to obviate the need to rely on primitive rods and clocks.

2.2.1 Bona fide chronometry

It is probably correct to endow Synge with the title of the most dedicated proponent of a (primitive) chronometric take on GR: in his seminal textbook (1960) he puts a primitive correspondence between (the general relativistic notion of) proper time and the reading of some sufficiently good theory-external clock (such as an atomic clock)—dubbed a 'standard clock'—

[21] Note that this point can at most be motivational: the Marzke–Wheeler construction presupposes Lorentzian spacetime.

centre stage for any empirical account of the metric field; length measurements are then derivative on time measurements under the assumption of a constant speed of light.[22] Now, it is with accepting clocks as theory-external objects in the interpretation of GR that EPS (and their followers) take issue. Notably though, there is another, in fact distinct, tradition of criticizing presupposed theoretical clocks in GR: its progenitor, Einstein, and (among others) Feigl more generally (see e.g. (Feigl, 1950)), demanded that a theory provide a description of its 'observation theory'—including e.g. its clock constructions—without, however, any demand for constructivism or constructive axiomatization more specifically. Carrier summarizes succinctly the distinction between the two traditions:

> The constitutive principles of constructive axiomatizations are (1) the methodological requirement of direct testability and (2) the semantical postulate of linearity. [...] EFC [Einstein–Feigl completeness], on the other hand, is guided (1) by the methodological idea of explanatory power (that is, the idea that a theory should explain a large range of phenomena in a precise fashion by invoking as few independent assumptions as possible) and (2) by a semantical account that allows for reciprocal clarification of concepts and theories. [. . .] Concerning the drawbacks of both approaches, note that the whole observation basis of a theory cannot be constructively axiomatized and that complete theories (if only in extreme cases) may suffer from test restrictions. (Carrier (1990, p. 391))[23]

As noted above, bona fide chronometry *qua* epistemology of spacetime is prima facie less ambitious than the causal-inertial method insofar as the import of an external, non-primitive element to GR ruins any prospects of an Einstein–Feigl completeness—the idea that the theory provides its own observational theory—from the start.[24, 25] Note, therefore, that chronometry—and,

[22] Again, it's not obvious how uncontroversial this assumption really is: see Asenjo and Hojman (2017); Menon et al. (2018); Linnemann and Read (2021a).

[23] In the same passage, Carrier emphasizes nicely the methodological advantages of the Einstein–Feigl tradition over that of EPS (and constructive axiomatization more generally).

[24] It has been said before that chronometry picks out time as preferred for epistemic purposes over space. Note that proper time is a preferred notion in general relativity from an observer's point of view over that of proper length: the former is defined along her whole worldline, whereas the latter is just a reasonable concept within a small spacelike neighbourhood intersecting her worldine alone.

[25] The chronometric approach is one fall-back option in all those spacetimes which cannot come out of a constructivist axiomatization, say when no light rays nor any other sense of field moving approximately close to the null cone are available. Note though that this does not entail yet that tracking is impossible in such spacetimes without the chronometric approach: the Einstein simultaneity condition via light signals is not the only sensible operationalist manner to arrive at an operationalist coordinate system for tracking. See for instance the supplements of Janis (2018) and references therein for how to set up a standard of synchrony from clock transport alone.

indeed, any theory which is not Einstein–Feigl complete—is susceptible to the charge of 'legitimizing sin', in Einstein's sense (again, see Giovanelli (2014) for further discussion).[26]

2.2.2 Weylian (inertial) chronometry

The EPS scheme uses a clock-based criterion ('no second clock effect') to turn a Weyl spacetime into a Lorentzian spacetime in the axiomatic process. We now review critically clock accounts for a Weyl spacetime context with regard to their applicability within the EPS scheme: such accounts have been developed with a view to filling this lacuna in the original EPS article.[27] We structure the discussion of clocks in the context of Weyl spacetimes along two traditions: one of operationalist constructions and one of operationalist test criteria of clocks. Before we can discuss such clock accounts, we have to clarify what it means to have a notion of time admissible to clock measurements in a Weylian context to begin with.

Proper time in Weyl spacetime

In this subsection, we introduce the (not very well-known) notion of proper time for Weyl spacetimes. A notion of proper time for Weyl spacetime cannot be defined in direct analogy to the worldline interval in Lorentzian spacetimes, as such an object would not be invariant under Weyl (i.e. spacetime-dependent scale) transformations. Following (Perlick (1987); Avalos et al. (2018)), a suitable notion of proper time in Weyl spacetime that is extensionally equivalent to the standard notion when restricted to Lorentzian spacetimes can, however, be set up as follows:

1. Consider the distinction between generic parameterized geodesic and parameterized geodesics proper (i.e. one for which the right hand side of the geodesic equation vanishes). The former differs from the latter insofar as the proportionality constant between the tangent vector and

[26] Note that Einstein–Feigl completeness accords with Stein's injunction to 'schematize the observer' in physical theories: see Stein (1994), and Curiel (2019) for further discussion. Arguably, however, Einstein–Feigl completeness is a stronger notion than that of schematizing the observer—for the latter mandates simply that a theory make contact with the empirical, with experimental measurement outcomes, etc. which of course can be true without that theory explicitly modelling the measurement process. For example, modelling measurements 'in' QFT often requires coupling one's quantum fields to some non-relativistic quantum mechanical degrees of freedom, thereby stepping outside the strict remit of that theory (Grimmer, 2023): this, we take it, is a case of schematizing the observer without Einstein–Feigl completeness.

[27] For further discussion of this issue, see Chapter 1.

its change relative to itself can be non-zero, i.e. $\frac{D\gamma'(u)}{du} = f(\gamma(u))\gamma'(u)$ with $f(\gamma(u))$ arbitrary for a generic parameterized geodesic, and $f(\gamma(u)) = 0$ for a geodesic. One might want simply to define a notion of proper time for geodesics in Weyl spacetimes as that parametrization of a generic parameterized geodesic that makes it a parameterized geodesic proper—but this notion cannot be directly taken over to arbitrary worldlines.

2. Importantly, the geodesic criterion for proper time $\frac{D\gamma'(u)}{du} = 0$ is equivalent to $g(\gamma'(\tau), \frac{D\gamma'(\tau)}{d\tau}) = 0$ where g is a representative of the Weyl metric. This orthogonality condition, can, in contrast to the proper time criterion based on the generic parameterized geodesic equation, be applied to timelike curves as well.
3. Thus, define a timelike curve in a Weyl spacetime as parametrized by proper time u for a segment between two events iff $\frac{D\gamma'}{du}$ is orthogonal to $\gamma'(u)$ for all u in that segment. (As stressed before, it can be shown explicitly that this notion of Weylian proper time reduces to the standard notions of proper time for Lorentzian and Weyl-integrable spacetimes more generally.)

Compare this to the definition of proper time for Weyl spacetime taken up by EPS, according to which 'a time-like curve γ is parameterized by proper time if the tangent vector γ' is congruent at each point of the curve to a non-null vector V which is parallel-transported along γ' (Avalos et al., 2018, p. 26), i.e. $g(\gamma', \gamma') = g(V, V)$. While it can be proved that this definition is equivalent to that of Perlick (see Avalos et al. (2018), proposition 2), Avalos et al. (2018) have a point in stressing that Perlick's presentation is more satisfactory, as it is a definition of proper time explicitly in terms of the worldline's properties alone—rather than relying on rather abstract 'comparison' vector fields.

We will now consider two traditions of clock considerations—one which aims at explicit clock constructions, and another which provides a test criterion for clocks.

The Langevin clock construction

Put simply, the idea is to construct a clock from light rays and test particles—i.e. types of idealized motion without back-reaction—that move back and forth in a well-defined segment and thereby read out a suitable (generalized) notion of proper time. For Lorentzian spacetimes, Fletcher (2013) provides a theorem on how light rays in a light clock moving on a time-like curve measure out some quantity proportional to proper time between two ticks (the 'accuracy' requirement), and do so in a regular fashion, i.e. the interval between two ticks is always the same (the 'regularity' requirement).[28] Correspondingly, for Weyl spacetime, Köhler (1978) has shown that light clocks moving along geodesics fulfil the regularity requirement.[29]

The problem with such clock constructions (and ultimately the reason why they do not immediately help in the EPS approach) is that they presuppose a notion of spatial regularity in order to latch onto the worldline they are supposed to measure out. In the Lorentzian case, this means that one needs to presuppose what is known as 'Born-rigidity' to hold between both sides of the clock. In the Weylian case, the problem is even more acute: first, there is no naïve sense of constant length available due to Weyl scaling. Second, note that the Weylian notion of proper time is non-geometric, i.e. it is not just an integral over local features but one over features that each depend on the starting position of the clock; after all, the explicit formula for the proper time between the parameter values t and t_0 of a curve in Weyl spacetime as discussed above is given by Avalos et al. (2018, p. 259):

$$\Delta\tau(t) = \frac{\frac{d\tau(t_0)}{dt}}{(-g(\gamma'(t_0), \gamma'(t_0)))^{\frac{1}{2}}} \int_{t_0}^{t} e^{-1/2 \int_{t_0}^{u} \omega(\gamma'(s))ds} (-g(\gamma'(u), \gamma'(u)))^{\frac{1}{2}} du. \quad (2.3)$$

A notion of proper *length* for Weyl spacetimes—obtainable via the presumption of constant speed of light—is thus presumably non-geometric (in the sense just defined above) as well. A third problem is arguably that mentioned by Delhom et al. (2020): no matter we know actually follows the timelike geodesic structure of a non-integrable Weylian spacetime (but only one of an integrable Weylian spacetime *qua* Lorentzian spacetime). But note the different status of this criticism, since it concerns the empirical status and not so much the mere conceptual status of such a clock. (In fact, one might take it that their point that matter is typically seen to fulfil the equivalence principle—thus ruling out a non-Weylian spacetime background for the matter—can be turned into a way of directly motivating a specifically Lorentzian spacetime geometry—see our argument to the effect of this at the end of 'Clock criterion'.)

[28] The tradition of rigorous light-clock constructions within general relativity goes back at least to Marzke and Wheeler (1964), who constructed a light clock for Lorentzian spacetimes, with the restriction that the light clocks move on geodesics; see also Desloge (1989). For some philosophical discussion of Fletcher's clock construction, see Menon et al. (2018).

[29] An alternative result for Langevin-type clock constructions in Weyl spacetimes is by Castagnino (1968), who for any torsion-free affine geometry (which includes any Weyl spacetimes) provides clock constructions. See also Delhom et al. (2020), who discuss the idea of a 'generalized' Märzke and Wheeler clock for non-metric (and in particular Weylian) spacetimes by focusing on the question of what kind of matter could realize such a clock.

Clock criterion

One might then want to relax the constructivist requirement and simply demand a criterion for distinguishing a good clock from a bad one. In fact, a clock criterion might under some circumstances still be seen as a part of a constructivist approach—albeit not of a linear but rather of an iterative constructivist approach (more on which below): start with some arbitrary Langevin setup, then tweak this 'clock' in some way: if the clock works better than before according to the criterion, then continue with that clock and tweak it again (by analogy with how actual precision clocks are set up, see (Tal (2016)). However, not only is convergence to a standard clock not guaranteed but even if the iterative procedure does converge, importantly, the EPS scheme will not be restored this way as a linear constructivist project.

In this subsection, we introduce Perlick's clock criterion for Weyl spacetimes which, using only facts about light propagation and freely-falling particles, allows us to test an arbitrary parameterization of a timelike curve to determine whether it can be associated with a standard clock. Following Perlick (1987, p. 1062), define a *clock* as follows:

Definition (Clock). *A smooth embedding $\gamma : t \mapsto \gamma(t)$ from a real interval into M such that the tangent vector $\dot{\gamma}(t)$ is everywhere timelike and future-pointing with respect to a Weyl metric g.*

The above definition captures the essential theoretical aspects of a clock—for any timelike curve may in principle represent the worldline of a clock; in particular, one may interpret the value of the parameter t as the reading of that clock. Nothing in this definition requires that the clock be a *good* clock—in the sense that nothing requires the ticking of the clock (i.e. the choice of parameterization) to be 'uniform', or to coincide with the regularities exhibited by material systems.

With this definition of a clock in hand, Perlick then uses the 'radar method' to 'spread time through space'—i.e. to assign coordinates *off* the worldline of the clock, at some point $q \in M$. This proceeds as follows (recall also our above discussion of the radar method *à la* Perlick). First, consider the curve γ representing this clock. One emits a light ray from an event on γ, say $\gamma(t_1)$; this light ray is reflected from q and is received back at some later point on γ, say $\gamma(t_2)$. (Note that this method presupposes that there is some means of reflecting the signal at q.) Then, the 'radar time' T and 'radar distance' R of q from γ are defined as follows:

$$T = \frac{1}{2}(t_2 + t_1), \tag{2.4}$$

$$R = \frac{1}{2}(t_2 - t_1). \tag{2.5}$$

Three assumptions here are worth stressing. First: we assume that the dynamics of the light signal isn't affected by the geometry of the spacetime under consideration—this, of course, is a non-trivial assumption. Second: this method of setting up radar coordinates presupposes the Einstein–Poincaré clock synchrony convention (see e.g. (Reichenbach, 1956)), insofar as q is assigned a temporal coordinate *half-way between* t_1 and t_2 on γ. This latter assumption is acknowledged explicitly by Perlick (1990, Section 2.1):

> Incidentally, if one replaces [(2.4)] by $T = pt_1 + (1 - p)\,t_2$ with any number p between 0 and 1, each hypersurface T = constant gets a conic singularity at the intersection point with γ. This clearly shows that the choice of the factor 1/2 is the most natural and the most convenient one. (If one allows for direction-dependent factors, one can get smooth hypersurfaces with factors other than 1/2. This idea, which however seems a little bit contrived, was worked out by Havas [(Havas (1987))] where the reader can find more on the 'conventionalism debate' around the factor 1/2.)

To be clear: the above more general choice for T, with direction-independent factors other than 1/2, goes back to Reichenbach, and is now known as the *Reichenbach–I synchrony convention.* On the other hand, a choice of direction-independent factors *à la* Havas is now known as the *Reichenbach–II synchrony convention.*[30]

The third assumption made above is that there exists a mirror at q capable of effecting the reflection—clearly, this is a substantive assumption, which will mandate prior coordination on the part of the experimentalist. (Note also that a mirror is not strictly necessary to set up these radar coordinates, and that any other method of ensuring that the signal returns from q to γ would suffice—e.g. a fly-by scenario where the incoming signal is crossed by another signal which effectively looks like a reflection of the incoming signal at a mirror.)[31]

[30] See Read (2023a, ch. 7) for further discussion, including how conic singularities can be avoided in the latter case.

[31] One would not need to make this assumption if one interpreted the radar coordinates in a modal sense—i.e. if there *were* a mirror at q then it *would* be reflected to γ. For more on such 'modal' constructivist approaches, see Chapter 3; for the time being, we leave it open whether appealing to modal notions is truly in the constructivist spirit. Note also that EPS themselves make liberal use of coordinates in their article—so they, too, seem to be invoking some degree of modal understanding of coordinates.

Now, Perlick (1987) shows that the radar method affords a means of ascertaining whether a given clock is a standard clock (with respect to some Weyl metric). The procedure works as follows. Consider a clock γ, and two freely-falling particles μ and $\bar{\mu}$ emitted from γ. The radar method assigns a time T and distance R to each event on μ, and a time $\bar{T}$ and distance $\bar{R}$ to each event on $\bar{\mu}$.[32] It can be shown that at time $t = t_0$ (which corresponds to $T = \bar{T} = t_0$)

$$\frac{\frac{d^2R}{dT^2}}{1-\left(\frac{dR}{dT}\right)^2}\Big|_{T=t_0} \pm \frac{\frac{d^2\bar{R}}{d\bar{T}^2}}{1-\left(\frac{d\bar{R}}{d\bar{T}}\right)^2}\Big|_{\bar{T}=t_0} = \left(\frac{dR}{dT} \pm \frac{d\bar{R}}{d\bar{T}}\right)\Big|_{T,\bar{T}=t_0} \frac{g\left(\gamma', \frac{D\gamma'}{dt}\right)}{g(\gamma',\gamma')}\Big|_{t=t_0}.$$

But this simply means that γ is a (Weylian) standard clock with respect to parameterisation t at time t_0 if and only if

$$\frac{\frac{d^2R}{dT^2}}{1-\left(\frac{dR}{dT}\right)^2}\Big|_{T=t_0} = \mp \frac{\frac{d^2\bar{R}}{d\bar{T}^2}}{1-\left(\frac{d\bar{R}}{d\bar{T}}\right)^2}\Big|_{\bar{T}=t_0}. \tag{2.6}$$

Suppose, then, that one experimentally ascertains that (2.6) obtains. The upshot is then that one obtains a criterion to test whether clocks in a Weyl spacetime are standard or not. With clocks certified as standard clocks in hand, one can then go and test whether there is a second clock effect— if not, the Weyl spacetime is integrable, and can indeed be replaced by a Lorentzian spacetime. To stress again: this is a *test* of whether a clock is 'standard', rather than an explicit clock *construction* (which was the topic of the previous subsection).

Notably, Hobson and Lasenby (2020, 2022) have recently called into doubt whether the generalised proper time for Weyl spacetime *à la* Perlick is physical. Specifically, they criticize the claim that the orthogonality condition presupposed in the derivation of the proper time formula, stated in coordinates as $g_{\mu\nu}u^\mu a^\nu = 0$, is required to transform non-physically with $u^\mu \to u^\mu$ under a Weyl transformation, when its actual transformation should be of form $u^\mu \to e^{-f}u^\mu$, where $e^f = \Omega^2$. The standard of physicality invoked by Hobson

[32] The radar coordinates by Perlick fix an event relative to the emission and reception times of one particle as well as angle values on that particle under which the probing signal was first emitted. Notably, though, the procedure for determining the parameter time difference between time of reception and time of emission as a 'standard clock' is completely independent of the angle information, and thus does not at all rely on recourse to structured devices such as sextants (more on which in Chapter 4), which would otherwise render the approach less powerful (one would thus perhaps not presuppose external clocks from the start but already quite problematically external angle-measuring devices, i.e. sextants). In other words, we can think of the emission and reception time used here (in order to define R and T) as two out of the four coordinates of the EPS coordinates.

and Lasenby is, however, debatable:[33] they consider observed quantities necessarily to be described locally by a tetrad system; the velocity measured by the observer is, for instance, supposed to be $u^a = e^a_\mu u^\mu$ (where a denotes the lab index and μ the usual spacetime index), and not u^μ; given that it is u^a that is taken to be physically relevant, it is also this velocity that has to be invariant under Weyl transformation (or so they claim). With e^a_μ transforming as $\exp(f)e^a_\mu$, it would indeed follow that u^μ transforms with $\exp f$, i.e. has 'Weyl weight' -1.

Interestingly, such an ascription makes the length of the tangent vector invariant under a Weyl transformation ($g_{\mu\nu}u^\mu u^\nu$ with $g_{\mu\nu}$ of Weyl weight $+2$ has a zero total Weyl weight). But—and hence the problem for Perlick's criterion—this means that the notion of proper time *à la* Perlick is no longer physically sensible, as it is no longer invariant under Weyl transformations (if $u^\mu := dx^\mu/d\xi$ has weight -1, so does $d/d\xi$ for any parameter (including proper time) ξ—x^μ after all does not change under Weyl transformation). More than that: (1) Massive particles are—since otherwise not in line with the scale-invariant background geometry—only instantiable through a dynamically coupling auxiliary field. (This requires making recourse to a Lagrangian structure for the particle and thus in a sense to at least placeholder relations to the dynamics already.) (2) The adequate connection is not the standard one as considered by EPS and the related literature but the fully covariant connection $\nabla*_\mu := \nabla_\mu + \omega B_\mu$ where ∇ denotes the standard Weyl connection, ω denotes the Weyl weight of the object to act on, and B_μ is the 'Weyl potential'. Interestingly, though, neither velocity nor acceleration of massive particles that are in free fall are generally parallel propagated relative to this fully covariant connection $\nabla*$.

It is arguably not immediate, though, that u^a is more physically relevant than u^μ and should thus count as invariant under Weyl transformations; after all, the tangent vector structure of the manifold—and not the local Minkowski structure associated with the tetrad formalism—may be seen as immediately physically relevant (after all, the metric and not the Einstein–Cartan formalism of GR is the default physical rendering of general relativity).[34] But as already pointed out, a mere clock criterion (such as that of Perlick) fails to offer a genuine constructivist link to Lorentzian spacetimes in any case.

The overall leap from Weylian to integrable Weyl/Lorentzian structure thus remains unsatisfactorily resolved when judged from the original constructivist

[33] Hobson and Laseby seem to presuppose a form of equivalence principle that would, however, require further discussion.

[34] Arguably, though, this is a weak rebuttal given that some matter (in particular, spinorial matter) requires tetrads for its formulation.

spirit of EPS. Efforts towards more constructivist alternatives to working via Perlick's clock criterion have been made either at the cost of the constructive nature of the scheme, i.e. the empirical immediateness of the axioms, by accepting (quantum-originating) matter waves (for instance—see (Audretsch and Lämmerzahl (1991))), or through switching altogether to an alternative (albeit related) scheme rather than EPS, which would, however, have to be scrutinized on its own merits (see (Schelb (1996a))). A criterion which requires the metric to show a certain symmetry, as already provided by (Schelb (1996a)), might give rise to the hope that a 'pure' constructivist criterion to differentiate integrable Weyl spacetimes from non-integrable spacetimes in terms of nothing other than radar coordinate events is possible (after all, metric entries do get described as measurable in the EPS scheme), but so far this approach has not been developed satisfactorily.

At the end, it is worth stressing that one might also take a route from Weylian spacetimes to integrable spacetimes via a (kinematically understood) weak equivalence principle[35] which is fully in line with the EPS constructivist spirit, or so we would argue. After all, one might simply add an additional empirical axiom to the effect that the projective structure is effectively zero for sufficiently small neighbourhoods—thereby immediately enforcing a Lorentzian geometry.

2.2.3 Parameter clocks

The EPS axiomatization in any case builds on clocks in a very limited sense, namely on what one might call 'parameter clocks' (as also noted in (Schelb 1996a, p. 1324)): some arbitrary parametrization of a wordline might be thought of as serving as a clock in the sense that any monotonically growing function of an object (like the height of a still-growing tree) can do so, even when there is no straightforward notion of periodicity associated with that object.

It should be clear that no objection against the clock-free approach can be mounted from the use of parameter clocks; their assumption commits one to nothing but a time-forward-moving process—in other words, the assumption that there is something like time for clocks to be measured to begin with.

To summarize this section, then: there is a (by now well-known) hole in the original EPS axiomatization, in the sense that EPS assume the absence

[35] See Linnemann et al. (2023a) for a discussion.

of a second clock effect in order to move from Weyl to Lorentzian spacetimes, but do not provide explicit clock constructions in Weyl spacetimes consistent with the axioms in order to underwrite this claim. Though efforts have been made to make good on such clock constructions in Weyl spacetimes, there remains work to be done on this front; moreover, Perlick's clock *criterion* clearly does not fully realize the constructivist's ambitions in this regard. Finally, EPS also make use in their construction of what we have called parameter clocks, but this should be regarded as being comparatively unproblematic.

2.3 Varieties of constructivism

We ascend now to a more general level. The purpose of this section is to characterize, at the greatest possible level of generality, the different varieties of constructivism. To this end, it is helpful to recall again Carrier's presentation of constructive axiomatics:

> The constitutive principles of constructive axiomatizations are (1) the methodological requirement of direct testability and (2) the semantical postulate of linearity. A theory is to be founded on propositions that are immediately amenable to experience; and the concepts employed by that theory should be clarified first (by relating them to experiences) and afterwards used to build up the theoretical edifice. The theory must not be used to elucidate the concepts; never turn your eyes back. (Carrier (1990), p. 391)

In this section, we argue that both putative constitutive principles (1) and (2) can be decisively relaxed; in this way, a richer understanding of constructive axiomatics, constructivism, and their limits, is uncovered.

2.3.1 Theoretical versus intuitive constructive approaches

Carrier's first constitutive principle of constructive axiomatizations—namely that axioms should be 'immediately amenable to experience'—leaves room for interpretation; what qualifies as 'immediately amenable to experience' is a function of context. To sharpen the discussion of Carrier's first constitutive principle, let us introduce the term 'observation theory' for any theory whose empirical content is known and itself taken to be relatively well accessible. (We have already seen something of the notion of an observation theory in our above discussion of Einstein–Feigl completeness.) With this in mind, we then propose a crude distinction between (i) 'intuitive constructive axiomatics', a constructive approach which makes no recognizable recourse to any (physical) observation theory—all initial empirical statements are intuitively given and formalized, and more advanced statements expressed in a linearly self-erected observational theory[36]—and (ii) 'theoretical constructive axiomatics', a constructive approach which (at most) makes recourse to (physical) observation theories that are related to the phenomena independently from the target theory.

The original EPS scheme, then, is largely an instance of (i). To see this, let us understand better how exactly observations are made according to EPS and how they are formalized into axioms. Now, the most basic observation adhered to in the EPS scheme is that the 'world' consists of events; strictly speaking, EPS should have made this explicit as their first observational axiom.[37] Then, EPS note that there are different kinds of sets in event space M corresponding to different histories of objects: one kind of set, which corresponds to the history of particles (simply called particles from then on), and another one, which corresponds to that of light rays (light rays from then on); again, the observation that there *are* particles and that there *are* light rays is suppressed as an explicit axiom of EPS. Once these two different types of event families are in place, one goes on to characterize observationally these two families of events (axioms D_1 and D_2) and ultimately—observing that these event families allow for coordinatizing other events (axiom D_3)—one can start expressing observations about individual events (which will ultimately enable the observation of projective and conformal structure). *At the general level*, then, it seems fair to say that EPS bootstrap their own observational language, and with it, their theory of interest (GR), from the rather immediately accessible notions of events, particles and light rays, and one's immediate control over them. This being said, it has to be stressed that even though EPS' axioms are generally ascriptions of formal properties based on observations, they also involve non-empirical idealizations. Take EPS' initial axioms for establishing the differential structure: that a sequence of events associated with a particle through time ('worldline') is ascribed the status of a smooth manifold does not follow from 'immediate experience'; clearly, smoothness is a technical idealization. Arguably, a naïve impression of continuity of the particle

[36] This is not to say that no non-physical theory, such as a logic or field of mathematics, is available and used.

[37] Arguably, EPS generally suppress any form of axioms that would amount to mere existence statements; rather, their axioms express observational attributions to already existing objects.

worldline (in the sense of a pencil line drawn in one go) may be something we can actually observe about particle worldlines. (Something analogous can be claimed about the smoothness/continuity of echos and messages.) Among the axioms on differential structure D_1–D_4, even the status of the more intricate axiom D_3 (postulating the existence of smooth radar charts on the set of events) seems to be in line with Carrier's characterization of the first principle of constructivism: when freed from the obviously technical (albeit not harmless) assumption of smoothness, it seems that we do observe light rays to pave event space in the way stated by L_1—which, recall from Chapter 1, logically precedes D_3—and thus ultimately by D_3.[38] The axioms, however, also partly contain conventions that could not in principle be tested. For example, the assumption in axiom D_4 that light rays are smooth is not a mere idealization, since *any* assumption about the one-way propagation of light is not amenable to empirical test (Salmon, 1977). Thus, one could say that, at least in some cases, the EPS axioms involve not merely idealizations, but also conventional assumptions.

Now, as no data is genuinely theory-free (Stanford, 2021), the distinction between intuitive and theoretical constructive axiomatics is of course somewhat vague (as we in a way already saw with the classification of EPS as an instance of (i)); strictly speaking, then, there can at best be a hierarchy of theoretical constructive axiomatics which differ from one another with respect to the degree of sophistication of their observational theories. This being said, it strikes us as fair and helpful to retain the bipartite distinction between intuitive and theoretical constructivist approaches (while recognizing its limitations), and we will continue to make use of this terminology in the remainder of this chapter. We give detailed examples of these two strands of constructive axiomatics below; before doing so, however, we consider the extent to which Carrier's second constitutive principle can be relaxed.

2.3.2 Linear versus iterative constructive approaches

The other central pillar of constructive axiomatics for Carrier—viz. the semantic postulate of linearity in the definition of theoretical terms—is also familiar from a more general project, namely Carnap's *Aufbau*—recall again the

[38] That said, note that we only observe this for a relatively sparse number of light rays in one small section of spacetime, so arguably there is something non-empirical going on in the generalization of this axiom to all of spacetime.

passage from the *Aufbau* given in §1.2. (See §2.4.9 for further discussion of the connections between the *Aufbau* and constructivism.) But in any case, a requirement of semantic linearity is at second glance unnecessarily restrictive: a proponent of constructive axiomatics might accept iterative definitions of terms instead of linear ones, at least under certain circumstances. Motivation in this direction can be drawn from the following paragraph from Carrier (who himself makes reference to similar ideas from Grünbaum (1973)):[39]

> One starts with an arbitrary geometry in the correction laws and determines physical geometry on its basis. The geometry obtained, however, does not generally coincide with the one used for the corrections. So in a second step, one employs the improved version of geometry to carry out the corrections and then repeats the whole procedure until an agreement is reached between the geometry entering the corrections and the geometry obtained by performing measurements with the accordingly corrected rods (compare Grünbaum 1973, p. 145). *If* this procedure of reciprocal adaptations indeed yields convergent results, the geometry that finally emerges is independent of the one used at the start and it is, furthermore, identical to the one obtained by exploring the coincidence behavior of transported rods in perturbation-free regions of space-time. So we can reasonably consider the resulting geometry as *the* geometry of space-time. (Carrier (1990), p. 385)

So in analogy, a constructivist—i.e. someone deserving of the title of a constructivist—could start with some previously known (empirically constructed) geometrical term that they are willing to leave 'free to flap in the breeze' in order to arrive at an even more convincing geometrical term; all that's assumed at the outset is immediately-given data (or data in terms of a prima facie acceptable observation theory), and that the initial geometrical proposal is in some weak sense based upon this data. For instance, in the flat-space spin-2 approach to GR, one starts out with a (local) Minkowski geometry together with a spin-2 field. From iterative correction procedures to the *dynamical* description of the spin-2 field, one then learns that the effectively resulting field replaces Minkowski spacetime as reference background geometry (or so the thought goes).[40] Generally, a theoretical constructivism could

[39] It is worth noting that, for instance, Carrier maintains that Grünbaum's specific iterative approach faces difficulties; however, these potential problems will not matter for the general conceptual points which we seek to make in this subsection.

[40] For a philosophical account of this, see Salimkhani (2020); see Linnemann et al. (2023b) for a critical stance on the 'spin-2 view'.

allow for an iterative development that starts with and develops its observational theory in a self-correcting iterative rather than a straightforwardly linear fashion.

At this point, it is also worth connecting this notion of iterative (theoretical) constructivism with that of a 'completionism' *à la* Einstein and Feigl, as introduced by (Carrier (1990)), and already discussed above. At a very coarse-grained level: Einstein–Feigl completionism (EFC) requires a theory to contain its own observation theory. Although there are clearly connections here with iterative constructivism, the latter differs from EFC in the epistemological maxim (Carrier (1990)): like the linear constructivist, the non-linear constructivist still starts out seeking to build up a theory from immediately accessible data through an accepted observational theory; but unlike the linear constructivist, she accepts that this might require a reciprocal approach. By contrast, the Einstein–Feigl completionist is motivated theoretically from the idea of minimal recourse to postulates external to the very theory under consideration in order to account for that theory's empirical predictions; for this they are, however, happy to assume the full theory from the start.

2.3.3 Examples

Given the above-introduced distinctions between intuitive and theoretical constructivism on the one hand, and linear and iterative constructivism on the other, we see now that there are four distinct strands of constructive axiomatics (see Table 2.1). We rehearse here a paradigmatic example for each type:

Linear-intuitive: An example of a linear-intuitive constructivist project is that of protophysics (as conceived and fostered by Lorenzen and Janich and their respective students in the late 1960s to early 1980s). The goal is to give a systematic, non-circular (thus linear) account of basic physical quantities such as distance, time, and mass in terms of simple, craftman-like operations in one's everyday

Table 2.1

	Intuitive	**Theoretical**
Linear	protophysics, EPS	EPS with quantum matter
Iterative	'inventing temperature'	spin-2 bootstrap of GR

Varieties of constructivism

'lifeworld' (*Lebenswelt*).[41] If successful, protophysics would provide a linearly derived basic theoretical structure relative to which empirical regularities can be formulated and judged on.[42] A second example of the linear-intuitive approach (and, indeed, axiomatization) is of course that of EPS, in which one begins with an intuitive observation theory in terms of light rays and freely-falling particles, and constructs therefrom more sophisticated spacetime geometries (Weyl geometries, and ultimately Lorentzian geometries).

Linear-theoretical: For an example of a linear-theoretical constructivist approach, consider the quantum mechanical generalizations of EPS developed by Lämmerzahl (1998). In these approaches, one begins with quantum mechanical waves (*qua* solutions of the Schrödinger equation), and again constructs spacetime geometric notions, and so forth. Although conceptually this approach is transparently linear, and in the same spirit as the original work of EPS, a key difference is that the initial observation theory in this latter case—viz. a theory of quantum mechanical waves—is substantially less 'intuitive' than that of the original EPS approach.[43] For this reason, it is reasonable to count it as a case of linear-theoretical constructive axiomatization.

Iterative-intuitive: An example of an iterative-intuitive constructivist project is that of the development of the field of thermometry, as presented in the book-length study by Chang (2004). According to this, one begins with elementary empirical observations, but then undertakes a 'self-improving spiral of quantification—starting with sensations, going through ordinal thermoscopes, and finally arriving at numerical thermometers' (Chang (2004, p. 221)). Such cases are also discussed—admittedly somewhat more abstractly—in (Read and Møller-Nielsen, 2020), in which it is argued that the relation between theory and observations can be understood within the framework of 'hermeneutic circle style reasoning' (cf. (van Fraassen (1980))), in which one begins with a basic observation theory, and on the basis of that theory constructs some more sophisticated theoretical edifice; that very edifice may, however, lead one to revise one's observation theory, at which point the process repeats.

Iterative-theoretical: An example of an iterative-theoretical constructivist project from physics practice is the famous spin-2 bootstrap of the Einstein field equations: the second-rank Lorentz-invariant tensor field $h_{\mu\nu}$ is postulated in a flat spacetime $\eta_{\mu\nu}$ such that it obeys the second-order equation of motion usually

[41] In fact, prior to grounding basic physical quantities, geometric notions and arguably even logical concepts have to be grounded linearly in everyday practice.

[42] The collection Böhme (1976) and the *Philosophia Naturalis* special issue on protophysics, edited by Janich and Tetens (1985), are good starting points to the literature. Major works include the protogeometric accounts by Inhetveen (1983), Lorenzen (2016), and Janich (1997), and the proto-physical account of time by Janich (2012).

[43] To some extent, it's hard to imagine an observation theory *less* intuitive, in light of the quantum mechanical measurement problem! In fact, it is even unclear that the Lämmerzahl construction really satisfies the desideratum of constructiveness above—that the (physical) observation theory (should be) strictly closer to the phenomena than the target theory. We will discuss these issues in much greater detail in Chapter 3.

associated with linearized gravity (called the 'Fierz–Pauli equation'). The central idea then is to allow for a sourcing of the Fierz–Pauli equation from matter energy–momentum content. As the resulting equation is, however, found to be inconsistent (hitting both sides of the matter-sourced Fierz–Pauli equation of motion with a divergence operator leaves the Fierz–Pauli part equal to zero while the divergence of the matter energy–momentum tensor is unequal to zero; if it were equal to zero as well, there would be no interaction with the Fierz-Pauli field h to begin with), one starts, as a remedy, to take into account the energy–momentum tensor associated to the h-field itself. Contrivedly, adding an energy–momentum tensor for h from the given equation of motion generates a new equation of motion for h whose energy–momentum tensor contribution to h has to be taken into account as well so that an iterative relation is generated. In the limit of infinite iterations, the Einstein field equations for a composite field $g_{\mu\nu} := \eta_{\mu\nu} + h_{\mu\nu}$ where η is the Minkowski metric, are claimed to arise.[44]

2.4 Constructive axiomatics in context

Having now (a) introduced some of the history of constructive axiomatics, (b) explained the connections between the EPS axiomatization and Weyl's theorem and presented some attempts to fill holes in the original EPS approach (§2.1), (c) considered the relations between constructive axiomatics and chronometry (§2.2), and (d) classified different varieties of constructive axiomatics (§2.3), we turn now to considering the connections between constructive axiomatics and various other projects and research programs in the foundations of spacetime theories.

2.4.1 Other constructive axiomatizations of spacetime theories

The literature includes various other constructive axiomatizations of spacetime theory (see Castagnino (1968); Hayashi and Shirafuji (1977); Majer and Schmidt (1994); Schelb (1996b); Schröter (1988); Schröter and Schelb (1992b); Schröter and Schelb (1992a)) that are, albeit less known, similar in spirit to EPS. On the other side of the spectrum, there are also various straightforwardly deductive axiomatizations (see Andréka et al. (2002); Benda (2008); Bunge (1967); Cocco and Babic (2021); Covarrubias (1993); Mould (1959) for some examples).

[44] Despite the seeming appearance even of a derivation, the inference suffers from several ambiguities. See, for instance, Padmanabhan (2008) and Linnemann et al. (2023b). A central criticism concerns ambiguities in setting up the consistency condition due to ambiguities in how to calculate the energy–momentum tensor for h. But see also Deser (2009) for a defence.

One case of a constructive axiomatization of a spacetime theory distinct from that of EPS but of particular interest in its own right is that of Hehl and Obukhov (2006)—what's known as the program of 'pre-metric electromagnetism', according to which certain pieces of spacetime structure (in particular, the metric signature) are to be built up (or, if one prefers, fixed via Lewisian functionalism *à la* Butterfield and Gomes (2023)) from elementary axioms about electromagnetism, formulated in terms of differential forms and *sans* metric field. For a recent philosophical analysis of that project, also engaging with the EPS axiomatization, see Chen and Read (2023). The main point which we wish to stress here is that the axiomatization of the pre-metric program seems to go beyond EPS insofar as it is able to derive *dynamics*, rather than just kinematical structures alone, from its axioms. On the other hand, the approach is unable to recover the full metric field; moreover, it counts as a more theoretical version of constructivism than EPS, insofar as (one might at least argue) axioms about electromagnetic fields are further removed from direct empirical experience than are those of EPS.

Interestingly, certain accounts cannot be easily filed as constructive or deductive. For example, the famous axiomatization of the special theory of relativity due to Robb (1914) takes as primitive a causal relation, which one may or may not take to be empirically basic/unproblematic. (For more on Robb's axiomsatization, the related 'causal theory of time', and their connections with EPS, see (Winnie (1977); Sklar (1977)).[45])

Another example in this direction is the account by Hardy (2018). Hardy argues for a broadly constructive approach to spacetime, not merely as a way of explaining classical spacetime but also as a promising route towards quantum gravity (see Chapter 3 for further discussion). He suggests that Einstein's original route to GR can be understood as a 'construction' based on seven principles (which single out the general framework for spacetime theories, i.e. the kinematics) and three additional elements (which single out general relativity, i.e. the dynamics). We may therefore compare the approach of Hardy (via Einstein) for constructing the kinematics to the EPS construction. In the schema of §2.3, Hardy's approach is clearly theoretical rather than intuitive

[45] Robb (1914), and later Winnie (1977), in his own modification of Robb's causal approach, used a definition of straight line which is materially adequate only in flat (Minkowski) spacetime. This is a substantive but unacknowledged assumption in their proofs of how to construct Minkowski spacetime structure from a casual compatibility relation alone, and serves to reconcile their approaches with the observation going back to Weyl (in his exchange with Reichenbach—again, see Rynasiewicz (2005) for the history)—that the structure of a Lorentzian manifold cannot be recovered from light cone structure alone. Our thanks to Adam Caulton for discussion on Robb's and Winnie's approaches.

since the seven principles don't describe immediate empirical facts but rather prescribe steps in setting up a coordinate system and putting fields on it—e.g. one of the principles is 'there is no global inertial frame,' which is not something that any local observer could hope to establish empirically (we might find local phenomena which we regard as evidence for the nonexistence of a global inertial frame, but we cannot make global observations and thus we cannot directly observe the existence or nonexistence of a global frame; and moreover it is unclear that 'nonexistence' can be subject to direct observation even in more ordinary cases). It is also structurally linear, in the sense that it involves no iterative steps, although one might contend that some of the principles seem to rely on specific knowledge about the target theory in a way which makes the approach somewhat less linear—e.g. one of the principles asserts that the theory should be defined in terms of local tensor fields based on the tangent space, which seems difficult to justify in terms which don't take account of background knowledge about general relativity.

Hardy goes on to propose a tentative 'constructive' approach to quantum gravity, with some of the details yet to be filled in. As in the case of Hardy's construction for GR, this approach is clearly theoretical rather than intuitive. Indeed, the axioms are necessarily very far removed from any possible empirical observations. For example, one axiom postulates the existence of 'indefinite causal structure,' which does not correspond to any known observation—and in fact it is quite difficult even to come up with a hypothetical empirical observation which could be regarded as instantiating indefinite causal structure. Given this feature, it's unclear whether we would want to regard Hardy's approach to quantum gravity as 'constructive' in the sense of EPS—perhaps it should simply be taken as a principle theory (more on which below) with non-empirical principles.[46]

2.4.2 Relationship to constructive mathematics

The constructivist aspect of EPS (and related programs mentioned above) shares motivations with constructive mathematics (which, in our nomenclature, would in fact better be named 'constructivist mathematics').[47] Both approaches regard non-constructivist procedures as being in some

[46] For more on constructive approaches to quantum gravity, see Chapter 3.

[47] Note, though, that EPS never mention—let alone endorse!—constructive mathematics. In our view, this is perfectly legitimate, in spite of constructive axiomatics (of physics) and constructive mathematics being natural bedfellows, as we explain below.

sense epistemically opaque *vis-à-vis* a constructivist approach. Constructive mathematics is therefore an instructive formal instance of constructivism that allows for appreciating the epistemological concerns at play in any constructivist effort.

Constructive mathematics is agnostic about the law of excluded middle—i.e. '$\forall p \in$ the set of Propositions: $p \vee \neg p$'—and thus dismisses it for all practical purposes, including as part of a proof technique. This scepticism towards the law of excluded middle is tantamount to scepticism towards proofs by contradiction of the form $\neg\neg p \rightarrow p$ (basically, a reformulation of the law of excluded middle). For the latter, it is quite clear why its assumption could be problematic: the proof-by-contradiction is indirect and thus opaque; rather, a proof should (or so a constructivist about mathematics would maintain) explicitly 'construct' what is supposed to be shown. Note that an actual dismissal of the law of excluded middle would also mean the dismissal of the axiom of choice (as the latter implies the former: see (Bauer, 2017) for a simple demonstration).

Now, if the usual repertoire of proof techniques becomes impoverished to the above degree, then it is not clear how much of standard mathematics (or something close to it) can be reproduced.[48] It was Bishop (1967) who for the first time demonstrated that a mathematical field—in his case that of analysis—could be given a viable constructivist reformulation. It is worth stressing that constructivism *à la* Bishop is of an epistemic character—it is motivated by a scepticism towards methodology. Brouwer's constructivism, known as 'intuitionism'—the first form of mathematical constructivism—arguably shares these concerns but the ultimate motivation is ontological: mathematics, a mind-dependent affair, needs to be studied by seeing how it is built upon a minimal ontology of intuitive statements (axioms). (See Bridges et al. (2022).)

If we now look at the case of spacetime constructivism, we see that there are clear parallels to epistemic constructive mathematics *à la* Bishop: rather than accepting the existence of some mathematical structure as empirically relevant just because it allows for deducing empirically testable results, basic experiences should—admittedly under some idealizations—allow for constructing the spacetime structure of interest. (Similarly, one could argue for an analogy to this effect between ontological constructive mathematics *à la* Brouwer and an ontologically oriented reading of spacetime constructivism, as discussed at

[48] Hilbert's lines of complaints in this direction are famous: 'Taking the principle of excluded middle from the mathematician would be the same, say, as proscribing the telescope to the astronomer or to the boxer the use of his fists' (Hilbert, 1927).

the end of Chapter 1.)[49] Also worthy of mention here is that a significant advantage of constructive mathematics is that the proofs are not merely existence or uniqueness proofs, but instead provide instructions for *how to construct the object in question*, which is sometimes a very useful and informative thing to be able to do. So arguably the motivation is not just epistemic but also pragmatic: here, too, there is an analogy with constructivism in spacetime theories, in the sense that the latter reveals the basic 'constituents' of the objects under consideration (e.g. projective and conformal structure composing a Lorentzian metric field), and how said constituents might also be arranged in different ways, in order to arrive at different structures (e.g. Finsler geometries, as we have seen).[50]

2.4.3 Constructivism and theory formulations

Famously, Einstein in 1919 drew a distinction between 'principle theories' and 'constructive theories':

> We can distinguish various kinds of theories in physics. Most of them are constructive. They attempt to build up a picture of the more complex phenomena out of the materials of a relatively simple formal scheme from which they start out. Thus the kinetic theory of gases seeks to reduce mechanical, thermal, and diffusional processes to movements of molecules—i.e., to build them up out of the hypothesis of molecular motion. When we say that we have succeeded in understanding a group of natural processes, we invariably mean that a constructive theory has been found which covers the processes in question. Along with this most important class of theories there exists a second, which I will call 'principle theories.' These employ the analytic, not the synthetic, method. The elements which form their basis and starting-point are not hypothetically constructed but empirically discovered ones, general characteristics of natural processes, principles that give rise to mathematically formulated criteria which the separate processes or the theoretical representations of them have to satisfy. Thus the science of thermodynamics seeks by analytical means to

[49] The aforementioned research programme of protophysics has been presented by its proponents (see e.g. Janich (2012)) as an extension of constructive mathematics into the regime of physics. (In that context, Janich does indeed use only constructive mathematics, or at least seeks to do so.) See our above discussion of protophysics as an intuitive constructivist approach to physics.

[50] To be clear: EPS do not *avoid* non-constructive proofs in mathematics; their proofs, indeed, are often non-constructive existence proofs. What we mean to stress here is only the conceptual affinity between constructive axiomatizations of physical theories and constructive mathematics.

> deduce necessary conditions, which separate events have to satisfy, from the universally experienced fact that perpetual motion is impossible. (Einstein, 1919)

Alongside phenomenological thermodynamics, Einstein identified his 1905 theory of special relativity as a principle theory. Now, while much has been written on the explanatory merits of constructive theories over principle theories (see e.g. Read (2020b)), what we wish to point out here is that there is a sense in which constructive axiomatizations—at least, intuitive constructive axiomatizations—are more akin to *principle* theories than to constructive theories. (Notably, the notion of 'constructive' in 'constructive theories' is fully independent of and not to be confused with how we defined 'constructive' in the context of 'constructive axiomatization' earlier.) The reason is that they precisely seek to build up the theories in question from empirically well-grounded axioms.[51]

Does this cast any aspersions over the constructivist project? In our view, it does not, for attempting to set any given theory on firm empirical footing is perfectly consistent with attempting to identify some deeper, physical explanation as to why that theory holds. Harvey Brown (p.c.) is suspicious of constructivist approaches such as EPS as a result of their being 'too operational' (see below for more on how to make sense of this claim); in our view, however, such suspicion is unwarranted, once one recognizes the compatibility of the two approaches as discussed above. It also deserves to be recognized that axiomatic approaches such as EPS seem to afford a means by which Brown's preferred program of the ontological reduction of spatiotemporal structure to material bodies can proceed—in this sense, he should also recognize the advantages of the approach (see below for further discussion).[52]

Moreover, the constructivist project may also be seen in light of a broader movement in physics towards taking principle theories more seriously. For example, Grinbaum (2007) argues that operational axiomatizations of quantum theory should be seen as principle theories, and argues that increasing interest among physicists in operational theories shows that principle theories

[51] From a systematic point of view, the principle/constructivist theory distinction is first of all an epistemic distinction in the sense of how we can learn about a theory of interest. Compare this to the framework/concrete dynamics distinction (as for instance pushed by Benitez (2019)), which is trying to distinguish the levels in abstraction between theoretical expressions and theories as such. In particular, a theory such as special relativity can be a principle theory, in the sense that it can be motivated phenomenologically, while at the same time star as the fundamental framework in a bottom-up model.

[52] We thank Harvey Brown and Chris Smeenk for discussion on these points.

are now being taken seriously as an end in and of themselves, rather than merely a step on the road to a constructive theory. Grinbaum (2017) interprets this trend as a move towards a form of idealism in which physics is understood to be primarily about language; Adlam (2022), on the other hand, interprets this trend within the realist tradition as a move away from object-oriented realism in favour of more structural realism. Both approaches agree that these operational axiomatizations have some advantages over the traditional constructive approach insofar as they offer a way of making sense of a theory which requires very few commitments to unobservable or theoretical entities—something which seems particularly valuable in the case of a theory like quantum mechanics, where the ontology of the theory is not at all transparent. More generally, similar claims can be made for other constructivist projects, including EPS: these 'principle theories' have value in their own right, and need not always be regarded as inferior to the complementary constructive theories.

The particular relevance of the EPS axiomatization in this context is that it emphasizes the fact that general relativity can be regarded as a principle theory, in much the same sense as SR is often acknowledged to be a principle theory.[53] Note that both SR and GR, *qua* principle theories, serve as *constraints* on possible dynamics for material fields (in the case of SR, that all material fields must have dynamics governed by Poincaré invariant laws; in the case of GR, that all material fields couple to a dynamical metric field). This is to be contrasted with thermodynamics, which is a purely phenomenological principle theory—there is no remaining freedom to postulate different dynamics consistent with the principles of the principle theory.

2.4.4 Constructivism and empirical interpretation

The EPS scheme (or constructive axiomatics more generally) can be seen as one specific way in which to address the interpretational problem for GR—i.e. the question of how to link its formalism to the empirical data and thus, in a minimal sense, to the world (see Carrier (2018); Dewar et al. (2022)).[54]

[53] The viewpoint of GR as a principle theory seems also in line with its thermodynamic–hydrodynamic interpretations, as put forward by Jacobson (1995), Padmanabhan (2012); Padmanabhan (2011), and also Hu (1999). Furthermore, even at the level of dynamics, it seems that there is phenomenological input into GR, say via the rather coarse-grained coupling in the Einstein equations of the metric field to matter in terms of energy–momentum alone.

[54] The question of whether general relativity needs an interpretation is also taken up by Belot (1996), Curiel (2009), and Linnemann (2021).

Whereas GR is standardly interpreted through fixed (often chronometric—see above) correspondence principles (see Synge (1959) and Malament (2012) for explications in this direction), the constructive axiomatic approach *à la* EPS generates incrementally ever more fine-grained correspondences, starting only with a basic representational ontology in terms of a set of events of which particles and light rays are supposed to be subsets.

Compare, for instance, Synge's presentation of the clock hypothesis as a correspondence principle to the way in which correspondences are built up by EPS. The clock hypothesis of general relativity equates the worldline interval length of a point particle trajectory to the actual proper time experienced by that particle; it becomes an interpretational principle at latest once the particle is understood as a stand-in for an idealized clock. By contrast, the constructive approach starts out by linking unspecified objects (set of events, particles and light rays) to the world which get rendered in a more and more detailed fashion via ascriptions (particles, for instance, are ascribed the status of smooth one-dimensional manifolds; furthermore, echos and messages defined between them are ascribed to be smooth, etc.); in the constructive approach, one thus dresses up an unspecified theoretical object which just acts, in a sense, as an 'anchoring' of the object in the world (say a set of events corresponding to a real-world 'particle'), with ever more detailed formal description. This is indeed in contrast to the standard chronometric coordinative definitions which associate 'finished products' in the theory to the world.

The lesson we learn is that, even though constructive axiomatics does not avoid the need for correspondence principles *per se*, it can take the weight from them by reducing them to a much weaker subset. After all, the usual concern with correspondence principles is that they are disturbingly brute stipulations—an impression which is, arguably, successfully weakened in constructive axiomatics.

Another (quite distinct) approach to the empirical interpretation of GR begins with the *local* validity of special relativity,[55] and bootstraps from this via certain additional principles to the kinematical structure of general relativity (these ideas are discussed by Brown and Read (2016), and more explicitly by Hetzroni and Read (2023)). This is in the spirit of a point raised by Lehmkuhl (2021), that the 'strong equivalence principle' (by which here is meant the assumption of the local validity of SR) can lead to the 'trickling up' of the interpretation of SR to GR. (Such ideas are also suggested in Brown (2005, ch. 9).) Immediately, one can see that such approaches are more theoretical

[55] See Linnemann et al. (2023a) for an explication.

(in our sense used above) than e.g. EPS, for they *begin* with the full theoretical edifice of special relativity. On the other hand, arguably one advantage of said approaches is that they thereby subsume *all* the empirical evidence which led to special relativity, before attempting to bootstrap to some new theory (namely, the general theory), rather than beginning with a very impoverished (and, for that reason, some might argue, less physical) set of starting axioms. Our conjecture is that it is this latter aspect which leads Brown to prefer such approaches over the EPS construction (as mentioned above).

2.4.5 Constructivism and operationalism

The EPS scheme has a strong operationalist flavour. It is important to distinguish, then, whether operationalism is meant here as a theory of *meaning*, or rather in a moderate fashion which stresses the need for operationalist *analysis*, rather than the absolute necessity of operationalist *definitions*. We take it that it is indeed not necessarily the case that the constructivist considers a theory-first approach as meaningless unless linked to the world by some operational interpretative rule. More likely, they will simply consider it less satisfactory than the epistemically less opaque route of linking the theoretical structures directly to experience (epistemically less opaque in the following sense: by choosing the empirical as the basis for the axiomatization, one obtains control and thus epistemic transparency with regard to which changes in the empirical impact which theoretical structures). Given that operationalism as a theory of meaning has become outdated (Chang, 2009), we will consider the moderate stance in the following.

To what extent, then, does the EPS scheme succeed with operationalist analyses? As we have seen above, the EPS scheme is limited to local neighbourhoods at least for non-static spacetimes. Even in the case of static spacetimes, any form of sensible tracking requires encodings within the signal (to showcase emission and receival times); from an operationalist point of view, the EPS scheme is in this respect really just schematic. In any case, it seems fair to see the EPS scheme as a form of a conceptual operationalism: one can lead back spacetime to manipulable operations with particle and light trajectories at least as a form of helpful mental picture.[56]

[56] Cf. Bridgeman's idea of mental, verbal, and/or paper-and-pencil operationalism—as opposed to that of a 'laboratory operationalism'. As Chang (2009) notes, Bridgman lamented that it was the 'most widespread misconception with regard to the operational technique' to think that it demanded that all concepts in physics must find their meaning only in terms of physical operations in the laboratory (Bridgman, 1938).

But more than that, the EPS approach does seem to provide a scheme which, once fleshed out through appropriate choice of particle and signalling items, promises to be physically realizable. The work of Audretsch and Lämmerzahl (1991) does replace the idea of particle by that of matter wave, but note that thereby just one still rather theoretical concept is swapped for another; even on this account, then, matter wave and light rays need to be fleshed out further by the experimenter. In a sense, none of this is of course news but rather part of the business of experimental physics when carrying out theoretical proposals in the concrete. Notably, quantum EPS—see the introduction of Chapter 3—may indeed not be regarded as an experimentally realizable scheme; in this sense, it is less operational (in a practical sense) than the classical EPS scheme.

2.4.6 Connections with geodesic theorems

In a typical textbook presentation of general relativity, the geodesic motion of test bodies is assumed—see e.g. Malament (2012, ch. 2). However, since (at least—see below) the work of Geroch and Jang (1975), authors have sought to *derive* the geodesic motion of small bodies from the Einstein equations in general relativity. (Actually, work to prove geodesic motion in general relativity—albeit of a very different kind to that of Geroch and Jang (1975)—goes back to Einstein and Grommer: see Tamir (2012) and Lehmkuhl (2017) for discussion.) Successor papers to Geroch and Jang (1975) include Ehlers and Geroch (2004)—in which back-reaction between the small body and the metric field is accounted for—and Geroch and Weatherall (2018)—which uses the machinery of 'distributions' in order to derive in addition that small *massless* bodies follow null geodesics. Insofar as one might then use these motions of small bodies in order to reconstruct the metric field via Weyl's theorem (Weyl, 1921), one might think that such geodesic theorems secure full access to the metrical structure of spacetime (something along these lines is suggested by Read (2020a)).

This reasoning, however, is somewhat confused. All geodesic theorems of the kind discussed above are proved from within the context of the *completed* theory of general relativity—thus, they have very little to do with the program of *constructive* axiomatics *à la* EPS. That being said, having (from within the completed theory) derived such motions, one could of course in turn apply the EPS machinery to these paths in order to re-derive the original metrical structure from which one began, as a kind of consistency check. Aside from this, however, it's not entirely natural—*pace* Read (2020a)—to situate such geodesic theorems alongside EPS, when the former begin with the very theory (the kinematics of) which EPS is designed to recover.

2.4.7 Relationship to conventionalism

Within the epistemology of geometry (on which see e.g. Dewar et al. (2022) for an introduction), constructive axiomatics (on which, as should by now be extremely clear, the kinematics structure of a theory is to be built up from elementary axioms which are supposed to have direct empirical significance) is sometimes set apart from conventionalism (according to which the structure of space and time is a conventional matter, which must be chosen on the basis of extra-empirical considerations), the latter being most famously associated with Poincaré (1902). What our discussions here make clear, though, is that constructive axiomatics *à la* EPS in fact goes hand-in-hand with conventionalism. For example: when EPS rule out Finsler geometries or torsionful geometries, they essentially do so by fiat. Moreover, we've already seen that some conventional stipulations (e.g. regarding freely falling particles) are retained in EPS, even on Coleman and Korté's modified version of the scheme.[57]

One central issue regarding conventionalism pertains to the extent to which the theory under consideration depends upon the assumptions made about the observation theory in use. As already insinuated, at a general level, independent of the specific case of the EPS scheme, there is a sense in which the linear constructivist approach can never fully succeed in building up theoretical structure 'from scratch': there are always conventional choices to be made, and the constructivist thereby only 'feigns' not to know what she is genuinely after. For instance, as Carrier (1990) notes, 'the EPS scheme does not relieve us from the need to decide about the presence or absence of universal forces'.

2.4.8 Constructivism and the dynamical approach

The dynamical approach to spacetime theories, promulgated by Brown (2005); Brown and Pooley (2001, 2006), and summarized recently by Brown and Read (2021), is an alternative to constructive axiomatics as an account of why the structure of spacetime is what it is (rather than otherwise)—this contrast was drawn recently by Dewar et al. (2022). One of the core tenets of the dynamical approach, at least in the context of theories with fixed spacetime structure such as special relativity or Newtonian gravity, is that spacetime structure *just is* a codification of the symmetries of the

[57] In this respect, constructive axiomatics is distinct from protophysics, the proponents of which claim to be able to circumvent all matters of conventionalism: see Dewar et al. (2022).

dynamical equations governing material fields, so that ultimately the nature of spacetime is to be explained by appeal to features of the dynamical laws.

Now, the constructive axiomatic approach might initially seem very different in spirit to the dynamical approach, insofar as the EPS construction (say) proceeds entirely from empirically observed motions without invoking any equations of motion at all (cf. our discussion in the final paragraph of §2.4.4). But following Anandan (1997), it is possible to see connections between the approaches—because after all, symmetry groups must also have something to do with the empirically observed motions, in which case it must be possible to extract the symmetry groups from those motions.[58]

In particular, Anandan notes that in a sufficiently small region,[59] the affine structure as defined early on in the EPS construction has as its symmetry group the affine group generated by the general linear transformations and translations in a four-dimensional real vector space. This affine group has as subgroups the inhomogeneous Galilei group and the Poincaré group—the former corresponding to non-relativistic physics and the latter to relativistic physics. Moreover, focusing on the Poincaré group, it in turn has as subgroups the translational subgroup and the Lorentz subgroup. The former acts on a small region around a given point and determines the projective structure, with the latter leaving invariant the null cone at each point and hence dictating the conformal structure; the relationship between these two spacetime structures can be understood in terms of the relationships between these two subgroups of the Poincaré group. The route taken by EPS can now be understood in something like the language of the dynamical approach. Thus, there is in fact a reasonably close correspondence between the dynamical understanding of spacetime in terms of symmetries and the constructive approach in terms of the behaviour of particles and rays. Indeed, in a sense the explanations of relativity offered by the dynamical approach can be regarded as simply a reformulation of the explanations offered by the constructive approach, arising naturally when

[58] We recognize that the symmetries of solutions of equations need not be the symmetries of those equations themselves—see Read and Cheng (2022); Murgueitio Ramírez (2024) for discussions of this point in something like this context. Nevertheless, it is surely true that the behaviour of material bodies, described by a particular solution to a particular equation, should in general have *something* to do with the symmetries of that equation, even if only in the sense that when we consider the set of all possible solutions, we should expect that symmetries of 'most' or 'typical' solutions to be in alignment with symmetries of the laws. (Our thanks to an anonymous reader for discussion on this point.)

[59] More formally, one might want to understand the relevant symmetry groups in the following in terms of approximate Killing symmetries, as done by Harte (2008) (see Linnemann et al. (2023a) for a philosophical discussion).

we abstract the symmetry groups away from the behaviour of test particles and arrive at spacetime on that basis. (This being said, a proponent of the dynamical approach might still complain that the EPS methodology *qua* route into GR does not begin with sufficiently rich theoretical structure, and in this sense remains 'too operational'—recall again our discussion in §2.4.4.)

Moreover, recall that in the original EPS construction, in order to arrive at Lorentzian geometry from Weyl geometry, EPS must invoke a somewhat ad hoc requirement: the stipulation that there should be no 'second-clock effect'; i.e. that parallel-transported vectors should not change in length. But Anandan contends that a less ad hoc approach emerges naturally within the dynamical picture if particles are replaced with quantum matter ones, subject to the requirement that the waves approach particle-like behaviour in the geometric optical limit.[60] For quantum waves have a natural frequency given by $mc^2 = \hbar\omega$, and thus the phase operator of such a wave can be used as a clock. In particular, we may imagine two such waves travelling from a common origin to a common destination; the metric along each path is determined by the Casimir operator m^2, but meanwhile the gravitational phase operator which generates the evolution along the path commutes with the Casimir operator, and therefore the Casimir operator remains the same as it is transported along each path, which means that the clocks' rates must agree again when they meet. Thus, in the quantum context the 'no second-clock effect' axiom can be regarded as following directly from fundamental dynamical symmetries.[61]

This is interesting for several reasons. First, it gives us a tantalizing glimpse of a deep underlying connection between quantum mechanics and classical general relativity: perhaps the reason EPS could not do without the ad hoc 'second-clock' postulate is because in order to fully understand and rationalize the nature of classical spacetime one must take into account its quantum underpinnings.[62] Second, it suggests that, *contra* the approach of EPS, it may in fact be necessary to invoke some dynamical considerations in order to arrive at the structure of spacetime: it seems that kinematics and dynamics

[60] Cf. Audretsch and Lämmerzahl (1991).

[61] Note that this is not entirely in the spirit of EPS, for the clocks here are not built up from basic, empirically informed axioms, but instead are identified in the structure of material (quantum) systems. (For another approach to building idealized clocks from quantum systems, see (Livine, 2022).) Nevertheless, this goes some way to showing that such clocks need not be assumed as primitive (which was the original concern about EPS' invocation of 'no second clock effect'). Still, one might take issue with the fact that one in a way here just *stipulates* the existence of a specific clock thereby after all (even though one explicates what kind of material fields allow for realizing it).

[62] This we take to be in the spirit of our Chapter 3.

are more closely intertwined than one might initially have suspected, and perhaps a complete understanding of the nature of spacetime depends on an understanding of the relationship between them. Of course, this latter lesson is in the spirit of Brown (2005).

2.4.9 Constructivism and *Der Aufbau*

As already alluded to by borrowing Carnap's definition of 'construction' for our own purposes, EPS' undertaking resembles Carnap's project in *Der logische Aufbau der Welt.* In general, it seems worthwhile to relate wider-ranging constructivist projects *à la* Carnap to domain-specific (i.e. physics-specific) constructivist accounts such as EPS: this task we undertake in the present subsection.

In *Der Aufbau,* Carnap considers hierarchical layered systems of statements—'constitution systems' (traditionally translated as 'constructions'!) —in which the vocabulary for statements in higher layers can be defined in terms of the vocabulary for statements in the next-lower layer and in which the lowest layer's vocabulary is called a 'basis'. The ambitious claim is then that the world as such can be expressed in terms of various constitution systems, say by using a physical basis, i.e. basic physical facts, a hetero-psychological basis, i.e. basic facts of observers' experience, or an auto-psychological basis, i.e. basic facts about one observer's very own experience. Importantly, Carnap stresses that the purpose of such reconstructions of the world in terms of constitution systems need not be epistemic; and that even though he himself takes the auto-psychological basis to be most fruitful when there is an epistemic interest behind setting up a constitution system, he explicitly leaves open whether such choice in the epistemic context can be overturned in the future.[63]

Despite such explicit qualifications, Carnap's work has in particular in the English-speaking world been read—or, rather, in absence of a proper translation, reported by Ayer, Quine, and others—as a project of reductionism in the British empiricists' tradition. However, a second major reading of Carnap's motivation has over time emerged—in the English-speaking literature in particular spearheaded by Friedman (1999)—which puts Carnap's intention much more into context with his phenomenological and neo-Kantian influences in Germany at the time of writing.[64] For Friedman, for instance, the

[63] Advances in neuropsychology (or so he speculates) might fancy a more naturalized approach again, and taking its basic vocabulary as a basis. See Leitgeb and Carus (2022, supplement A).

[64] For another reading, see Pincock (2005).

starting motivation to *Der Aufbau*, with its de facto one-sided commitment to the auto-psychological basis, lies in an attempt to bridge the gap between one's basic phenomenological impressions on the one hand and the objective physical world on the other, for the sake of structuring and solidifying the status of the phenomenological; it concerns the status of the phenomenological not of the outside world, or—more in the neo-Kantian spirit—the gap between the subject and the world as such. But as Friedman proceeds to argue, this original motivation ultimately gave way to the thought that the constitution systems can be read in metaphysical or epistemic dimensions at will as long as its content in a narrow sense is seen (and only seen) in structural relationships; the adherence to this predecessor of the tolerance principle then ultimately makes the phenomenological as well as the neo-Kantian motivation only one out of several possible readings (and also leaves room for the reading in terms of empiricist reductionism as an additional option).[65]

Now, *Der Aufbau*—no matter in which reading—is widely considered a failure. However, Leitgeb (2011)—who has set out to weaken and smoothen its claims to arrive at a more viable version—argues for reviving *Der Aufbau*'s project *qua* epistemological agenda in two different ways: (i) partly responding to the original empiricist motivations, an updated *Aufbau* can demonstrate a relevant part of (albeit not exhaust) the meaning of expressions. As Leitgeb details: 'if experience is understood in terms of a subjective basis that is relativized to a particular cognitive agent, then the so-determined empirical meanings may be considered to be among the internalist meaning components of linguistic expressions—the meaning components that are "in" this agent's mind—which are additional to externalist (referential) ones' (Leitgeb (2011, p. 270)). So, accepting with the mainstream that what is empirically accessible does not exhaust the meaning of theoretical terms, constructive axiomatics allows for working out the extent to which the individual observer can remark about the totality of meaning of theoretical terms (for the internal/external meaning distinction, cf. (Putnam (1981))). Secondly, partly responding to Carnap's original neo-Kantian motivation (at least on Friedman's reading), a new *Aufbau* may still be used to 'fill the gap between subjective experience and the intersubjective basis of scientific theories. After the protocol sentence debate in the early 1930s, philosophers of science more or less decided to conceive of the observational basis of science as being intersubjective right from the start; observation terms and observation sentences were meant to refer to

[65] See Leitgeb and Carus (2022, supplement A) for a detailed account of the *Aufbau*'s reception.

observable real-world objects and to their observable space-time properties' (Leitgeb (2011, p. 270)). In the case of EPS, one might then find justification in the project in that it helps to connect various observer viewpoints (you with your particles and light rays; me with mine), solidifying their viewpoints as parts of a single scientific outlook (that of GR).

Moreover, one might argue that a revived EPS constructivist project can be motivated for similar reasons: thanks to an EPS-like scheme, one learns about the operational content of projective/conformal geometry; and thanks to EPS, one is able to link the observer explicitly into the otherwise highly theoretical structure of the full kinematical picture. Notably, this second point is much stronger than a mere 'schematization of the observer' within a given theoretical structure (relatedly, see Stein, 1994; Curiel, 2019)—one reduces adherence to correspondence rules to a significant degree (in analogy to how relying on one's own protocol sentences, rather than relying on just anyone's protocol sentences, reduces adherence to correspondence rules to a significant degree).

2.4.10 Constructivism and constructive empiricism

The connections between constructive axiomatics and the constructive empiricism of van Fraassen (1980) are more than merely name-deep, and, indeed, are also worthy of some discussion. Recall that constructive empiricism is a thesis about the aims of science: according to the view, the aim of science is to construct theories which are adequate to all possible empirical observations—as opposed to construct theories which are literally true, *per* the scientific realist. For the constructive empiricist, the most epistemically modest stance to take with respect to propositions about the unobservable (which should be read as: unobservable *for us*) is agnosticism.

Insofar as propositions about (say) light rays and freely-falling particles are (let us grant here) empirically unproblematic, whereas propositions about (say) metric fields have to do with unobservables (as the typical adage runs: who has ever bumped into a metric field?), the constructive empiricist would be reasonable to treat the first set of propositions as literally true, but to remain agnostic about propositions in the second set. What constructive axiomatics reveals, however, is that perhaps the constructive empiricist can recover more than they might think vis-à-vis assent to propositions about the unobservable: if axioms about the empirically unproblematic (here, again, light rays and particles)—expressed as propositions—can be used to fix facts about unobservables (e.g. metric fields), then it seems that the constructive empiricist can—and, indeed, should!—assent to the literal truth about those latter propositions after all. In other words, by grounding propositions about theoretical structures such as metric fields in certain propositions about the empirical by way of the machinery of constructive axiomatics, the constructive empiricist can recover more than they might have initially thought about the unobservable world.

Now, insofar as we do not actually carry out all of the measurements needed to construct spacetime from empirical data, our claims about spacetime geometry seem to require some application of induction from actual measurements. Fair enough—but insofar as van Fraassen is willing to help himself to enumerative induction, we expect that he will find this acceptable. However, as is well-known, van Fraassen (1980) repudiates abduction (i.e. inference to the best explanation)—so a constructive empiricist in his mould should be sure that they do not justify (for example) the conventional inputs into the EPS axiomatization on abductive grounds.

2.5 Towards a restricted foundationalism?

We have noted before that a constructive-constructivist project *à la* EPS has some programmatic dependence (if not circularity) on the goal of its construction—viz. GR; arguably, it is thus not totally constructivist. However, it might even seem naïve to assume that the full content of a theory can be grasped constructively, i.e. from a non-theoretical or substantially less theoretical basis—we know by now, not just through the likes of Duhem, Neurath, Hanson, and Quine, that the theoretical is *generally* not exhausted by the empirical, and that (some sort of) coherentism has won over foundationalism. But this sentiment can be resisted: EPS—to all that we know by now—provides an exception: it is a partial empirical foundationalism for the kinematics of GR—even if it is not free of idealizations and imperfections.[66]

In this section, we, however, want to elaborate on what 'restricted' significance can be given to constructive-constructivist takes on physical theories, even if one wants to maintain in the context of EPS, as well as with the general consensus, that empirical foundationalism—to which constructive-constructivist programs first of all seem to aspire—fails.

[66] The dynamic of GR as well of that of matter fields could be built on top of the kinematics; however, as already stressed, the EPS axiomatization does not afford the resources to fix *specific* dynamics.

We already outlined various natural readings of EPS at the end of §1.7.3, each of which are different from empirical foundationalism in any case—namely, that of EPS (1) as a what-empirical-structure-is-sufficient explanation of kinematical structure, (2) as a how-possible explanation of theory change, and (3) as an ontological account of spacetime structure. Now, readings (1) and (2) seem to remain sensible readings just as before. Firstly, reading (1) does not become moot just because some starting assumptions are theoretical: the EPS scheme still allows for a counterfactual explanation of isolated theoretical structure in terms of basic observations of the agent even if some theoretical (background) assumptions need to be made. Secondly, reading (2) is a high-level heuristic understanding of EPS in any case. However, reading (3)—according to which EPS is about grounding spacetime structure in empirical posits and empirical posits alone—does not lend itself to a natural milder variation that is compatible with the failure of empirical foundationalism; or so one would think given that the original idea behind (3) that the basic ontology—and thus the basic explanations—should be grounded in the empirical and not in the theoretical (which would be the usual) only seems attractive as long as this sort of epistemological-turned-ontological foundationalism holds strictly. This being said, a more pragmatically-oriented ontologist, who aims at helpfully sorting nature but is not necessarily committed to certain metametaphysical convictions (such as that the empirical is standardly to be grounded in the theoretical), could still beg to differ.

Let us propose a different take amounting to a fourth reading which seems particularly suited to the idea of 'restricted foundationalism' (and in continuity with reading (1)):

> **EPS as a non-exhaustive account of the empirical meaning of theoretical terms relative to an agent:** The constructivist-constructive axiomatisation of EPS allows for taking an observer-first view—even if it cannot provide a theory-free construction from that viewpoint.

Quite concretely, EPS provides an 'observer-first' take on theoretical terms such as projective, conformal, and even metric structure. We take this reading to be ultimately the least controversial one (arguably, together with (1)).

We make another concrete case for the merits of restricted foundationalism in Chapter 3: we show how an EPS-like scheme can be used to provide a novel understanding of 'quantum spacetime', namely in terms of quantum signals: issues of quantum superposition of spacetimes (including the possibility of quantum diffeormophisms[67]) are led back to issues of how to think of quantum signals; this is advantageous to the extent that we have better intuitions for sensible options of how superposed signals can behave than for how superposed spacetimes can behave. In this sense, we thus defend the idea of

Quantum EPS as bridging the gap between classical and quantum spacetime: The constructivist-constructive axiomatisation of EPS allows for taking an observer-first view.

All achievements by EPS on the conceptual level (say, how to understand spacetime from the observer's eye) might still be discarded by some as mere reformulations of previous insights; in the case of quantum EPS, however, it is evident that spacetime superpositions have issues and that we make a leap forward in understanding these issues (or even identifying some of them to begin with) thanks to the constructive-constructivist approach.

For completeness, it is also worth pointing out that a constructivist mindset can usefully be put to action even across theories rather than between a theory and (parts of) its observational basis—and thus also independently of whether the construction basis is constructive in the narrow sense or that of Reichenbach or instead just admissible in some wider fashion. For this, consider the issue of spacetime emergence in GR from theories of quantum gravity: let us agree for a moment that quantum gravity approaches are indeed non-spatiotemporal in some relevant sense (see Le Bihan and Linnemann (2019); Linnemann (2021) for a dissenting view, and see Jaksland and Salimkhani (2021) for a pertinent critique of the loose usage of the words 'spacetime' and 'emergence' in spacetime emergence claims). The question that has kept people busy in the philosophy of quantum gravity community, then, is: how can spatiotemporal structure arise from non-spatiotemporal structure? Could it not be that the whole empirical success of standard physics—based on measurements in space and time—is undermined as long as it is not clear that spacetime exists fundamentally? (Huggett and Wüthrich, 2013)

Spacetime functionalists in this context (e.g. Lam and Wüthrich (2018, 2021); Huggett and Wüthrich (2021)) argue that the physical salience of these (presumably) non-spatiotemporal structures can be vindicated by showing how that non-spatiotemporal structure manages to functionally realize

[67] Relative to the wavefunction expressing the superposition of spacetimes, a 'quantum diffeomorphism' is the simultaneous application of individual diffeomorphisms to the different branches of the wavefunction (and the associated manifolds). The notion is discussed further in Chapter 3.

spatiotemporal roles. But note: a functional reduction (as with any reduction) first of all explains the reduced by the reducing. In this case, this means that the non-spatiotemporal explains the spatiotemporal, and not vice versa; in particular, the physically salient spatiotemporal structures can then not be adhered to for explaining the non-spatiotemporal structures. The proponents of spacetime functionalism claim, however, that the vindication of physical salience through a (functional) reductionist scheme runs both ways, that is from the reducing to the reduced but also from the reduced to the reducing; it is the latter which Huggett and Wüthrich (2021) rely on. As an intuition pump for this view, Huggett and Wüthrich (2021) bring up the example of the physicalist's reduction of mental states: they take it to be immediately plausible that the physicalist's functional reduction of pain in terms of brain state configurations can not only be read as strengthening the status of pain in the face of the physicalist's concerns. Rather, the physicalist's reduction can also strengthen the status of brain states in the face of the phenomenalist's worries. Fair enough, but this leaves room for concern, as the intuition here is a very different one from that of the constructivist: from the constructivist point of view, a physicalist reduction of pain bridges the gap between a physicalist understanding and a phenomenalist understanding (the physicalist can understand, i.e. can model the phenomenalist in her language). However, only a phenomenalist reduction of mental states would decisively bridge the gap *given* a phenomenalist basis to a physicalist worldview (the phenomenalist would understand, i.e. could model—at least to some extent—the physicalist's renderings in her language).

The constructivist who criticizes spacetime functionalism at the same time can make the positive point that the issue of the conceptual gap may just be addressable in the way we get to typical quantum theories of gravity. Take loop quantum gravity (LQG): LQG is constructed by a quantization scheme from general relativity. Surely, there are ambiguities and thus choices to be made on the road (see Linnemann (2022))—but at no point is it unintelligible how LQG arises *from the viewpoint of* GR. And that addresses the worry that it is hard to link up the spatiotemporal to the non-spatiotemporal: we should be concerned with linking from the spatiotemporal to the non-spatiotemporal because it is the *spatiotemporal*, not the *non*-spatiotemporal, that we can understand.

3
A Constructive Axiomatic Approach to Quantum Spacetime

Introduction

As we know well by now, the EPS axiomatization of (the kinematics of) the theory of general relativity (GR) purports to build up the spacetime structure of that theory from only (supposedly) indubitable empirical posits regarding light rays and particle trajectories. The EPS scheme and its subsequent amendments constitute invaluable tools for assessing the classical theory of spacetime, including the necessity and/or sufficiency of GR to account for a certain body of empirical data.

In this chapter, we aim to understand the extent to which something resembling the EPS approach to constructing spacetime can be applied when the inputs are quantum mechanical rather than classical (as in the original EPS construction). That is, we consider the EPS axiomatic approach to GR's spacetime structure with all classical light ray signals replaced by quantum light signals, and all particle signals replaced by quantum particles; this can be done, or so we will argue, in a natural fashion (indeed, a number of different natural fashions).[1] In making these substitutions and applying the EPS approach, one ultimately derives a superposition of metric structures as the relevant kinematical structure for quantum spacetime; moreover, as we will see, there is a way of interpreting these outputs in terms of branching spacetime structures.[2]

In more detail, what will the resulting kinematics look like? First off, it seems likely that we will not thereby arrive at a single Hausdorff manifold. This follows directly from a standard result of differential geometry: for a Hausdorff

[1] Notably, Audretsch and Lämmerzahl (1991) have extended the EPS scheme by considering matter that is explicitly modelled as the classical limit of quantum matter (postulates 1 and 2), and even as rays superposed to wave packages (postulate 2'). However, the possibility for the geometric background to become superposed in virtue of such superposition is not considered.

[2] One sometimes encounters objections to the notion of a superposition of spacetimes, on the grounds that Lorentzian metrics—unlike Riemannian metrics—do not form a vector space. It is by now fairly well recognized that this objection is a red herring: among other things, why think that $|\psi\rangle = |g\rangle + |g'\rangle$ should be such that $\psi = g + g'$ with ψ Lorentzian?

Constructive Axiomatics for Spacetime Physics. Emily Adlam, Niels Linnemann, and James Read, Oxford University Press. © Emily Adlam, Niels Linnemann, and James Read (2025). DOI: 10.1093/9780198922391.003.0004

manifold M, a geodesic beginning at a point p in M with an initial tangent vector x must be unique for a non-zero length of time t. This fails to be true in our case, because the expectation is that light rays will participate in spatial superpositions—for example, in a beam-splitter a light ray may be placed in a superposition of two different paths. Assuming that no spacetime point exists in more than one branch (we will return to this point), the two geodesics must lie along different points, so the geodesic from the divergence point fails to be unique.

There are a number of different issues to be considered here. First, there is a distinction between 'local' branching (in which curves 'split') and 'global' branching (in which curves do not 'split', but the spacetime in question has 'trousers'). It seems most natural to involve quantum superpositions of the kind described above in terms of local branching, but we might also have to take account of global branching. Second, there are various different mathematical ways in which both local and global branching could be modelled, and we will see later that each of these has its own advantages and disadvantages. What branching actually occurs will depend on the actual dynamics, so the goal here is to consider what kinds of mathematical structures will suffice to model all kinematically possible superpositions.

Difficulties arise because in the quantum case it is clear that axioms can't be *wholly* based upon empirical observations, as is purportedly the case in the original EPS construction. This is because in the case where light enters a superposition, we can't directly observe the light going down both paths (either because doing so would (effectively) collapse the wavefunction, or because in an Everettian picture observers are confined to individual branches of the wavefunction), so the existence of the superpositions is an inference that we make to explain the observed interference effects, rather than a direct observation.[3] Moreover, we also expect that massive particles will participate in superpositions and as a result we will get superpositions of different spacetime structures (whether at the level of the metric or of the manifold—we will return to this). Once again, these superpositions will not be observable directly, and in fact at present we do not even have indirect evidence for them, since no experimental demonstration of the existence of

[3] Admittedly, already in classical EPS one explicitly renders light and particle detection as indirect in the sense that one only measures the effect of light and particles rays on probing light signals in terms of the light signals' reception time. Quantum EPS' operational procedure of comparing light ray input to (interference) output is thus at least in this way continuous to the operational protocol in classical EPS.

spacetime superpositions has yet been obtained—the existence of spacetime superpositions is largely a theoretical conjecture at this stage.[4]

Against this, one might argue that the situation is not so different from the classical case—for the classical version of EPS takes for granted that operational axioms based on generalizations of our own direct experiences can be expected to hold throughout all of spacetime, yet evidently we cannot verify in any direct way that the axioms really do hold in regions of space very far from the Earth or very far in the past or future, and indeed, we probably cannot verify even indirectly that they really hold outside of the observable universe. Thus EPS seem to be implicitly relying on some kind of induction principle when they assume that operational axioms based on our direct experience can describe the whole of spacetime. And in a similar way, one might think that we can appeal to a principle of induction to argue that there is presumably nothing special about our branch of the wavefunction.

However, there are additional complexities in the quantum case. First of all, quantum mechanics tells us that the spatiotemporal experiences on which the EPS axioms are based have all taken place in a quasi-classical macroscopic reality stabilized by decoherence effects. So it may be reasonable to expect that similar operational axioms would hold in all branches of the wave function when we are dealing with spacetime superpositions involving large objects, since then decoherence will take effect and we can expect to have a stable quasi-classical reality in each branch. However, that leaves an open question about how we should think about relatively small-scale superpositions of spacetime; it seems reasonable to think that there could be objects massive enough that superpositions of them would come along with superpositions of genuinely distinct spacetimes but nonetheless still small enough that coherence could be preserved, but it does not seem obvious that we should expect operational EPS-style axioms to hold within each of these individual branches. This is related to ongoing discussions around spacetime emergence in quantum gravity: perhaps the operationally accessible features of spacetime only emerge in the decoherence limit, in which case we shouldn't expect operationally-defined axioms to hold even for small-scale superpositions. So the axioms that we employ in our quantum EPS construction will be somewhat conjectural, meaning that we have a certain amount of freedom about how to select them. The methodology we employ here will be based on keeping the

[4] That said, recently proposed experiments in table top quantum gravity in reach of actual implementation are seen by many to have the potential to change this. See Huggett et al. (2023) for a critical review.

axioms as close as possible to the original EPS axioms while reflecting standard principles concerning the behaviour of light and particles in a quantum world.

Now, one might ask why it is useful to carry out an EPS construction when the original operational motivations no longer apply. However, there is more than one way to motivate the EPS construction. One could see it—arguably as EPS themselves would have done—as a minimal empirically driven construction demonstrating that the kinematics of GR (or whatever spacetime theory at which one arrives via the construction) are the correct ones based on the existing evidence. Alternatively, one could take an essentially relationalist position which insists that the structure of spacetime is literally built up out of the behaviour of light rays and particles, and therefore if light rays and particles necessarily behave as specified in the EPS axioms, then it follows that spacetime necessarily has the structure that EPS derive. The EPS construction is then to be understood not merely as a *epistemic justification* for the relativistic kinematics, but at the same time seen as a metaphysical *explanation*, i.e. grounding for it.[5] But once we are aiming for explanation *rather* than epistemic justification in the quantum case, we are perfectly entitled to even use axioms describing behaviour that can't be directly observed—obviously the plausibility of the resulting explanation will depend on both the a priori plausibility of the axioms as well as the explanatory virtues of the resulting explanation, but then that is a common feature of scientific explanations. Indeed from this point of view it's quite remarkable that the original EPS construction apparently succeeds in explaining the nature of the relativistic kinematics based solely on axioms which can be verified by direct empirical observations (though it should be noted that there are disagreements about the extent to which it does indeed succeed—see Chapters 1 and 2). Moreover, in addition to the idea that one can use EPS to realize a relationalist vision with respect to the *metrical* structure of spacetime by grounding this in the existence and behaviour of light rays and particles, we note that if the radar coordinate construction is taken seriously and is understood to pertain to the behaviour of actual (and not merely hypothetical) light rays and particles, then it can be regarded as an implementation of a certain kind of structuralism (cf. (Esfeld and Lam (2008))) whereby spacetime points have their identities only in virtue of the behaviour of matter. The EPS construction can then be regarded as evidence that, at least at the level of local kinematics,

[5] For a similar case where the epistemic basis also serves as the metaphysical (or modal) basis, see Adlam (2022). Forms of Humeanism with an 'epistemic basis' arguably also are examples of this kind.

GR is compatible with such versions of relationalism/structuralism, provided that we supply ourselves with a sufficient number of light rays and particles. Attempting a similar construction in the quantum case thus offers interesting insight into how the relationalist/structuralist might fare in a quantum gravitational context and whether there are any additional challenges over and above those faced in classical GR.

In any case, quantum EPS has significance beyond metaphysics: attempting to perform a quantum EPS construction is an edifying educational exercise which confronts one with a number of questions regarding the nature and consequences of superpositions of spacetimes, with important lessons for quantum gravity research, in particular the right quantization of GR.[6] Thus this construction offers a different perspective on some of the conceptual challenges associated with quantum gravity, where we arrive at these problems from a top-down operational perspective rather than a bottom-up perspective based on some specific theoretical framework. Any bottom-up approach to quantum gravity must ultimately take a stance on the questions we will address, either by building a methodology based on certain answers to these questions or by deriving answers to these questions from its basic methodology. For example, the first question we have to answer in a quantum EPS construction is 'Can there exist superpositions of different spacetime structures?' and of course all mainstream approaches to quantum gravity (excluding semiclassical gravity) answer this question in the affirmative. The EPS approach offers us the opportunity to parse questions of this kind and examine the consequences of various possible answers to them in a way that is independent of any particular theoretical approach to quantum gravity.

We proceed as follows. §3.1 rehearses the relevant core features of classical EPS (for a more detailed introduction, of course, see Chapter 1). In §3.2, we consider the core obstacle to quantum EPS: how to relate spacetime structure across superpositions? The discussion leads to various ways to undertake a quantum EPS project, which will be presented in more detail in §3.3. In §3.4, we select two variants and exhibit some possible axioms for a quantum EPS scheme. Specifically, we envision one approach to EPS in which rays and particles themselves are quantized by defining 'events' on a Hilbert space and then defining rays and particles as subsets of the set of events, such that each one can be written as a superposition of smooth one-dimensional manifolds, giving rise to a superposition of spacetimes in which these different manifolds

[6] As we have seen in the previous two chapters, one can also give EPS an epistemological emphasis, insofar as it can be used as a tool for the *discovery* of certain spacetime structures.

live. Separately, we envision an approach to EPS which makes use of non-Hausdorff spacetimes to allow branching, so a particle is defined as a smooth one-dimensional non-Hausdorff spacetime, so the particle can have branching trajectories which will ultimately give rise to a branching spacetime.

Finally, in §3.5, we draw the following major lessons from this work:

- *Physical lesson:* From a strict relationalist point of view, spacetime superpositions may not be so conceptually distinct from spatial superpositions.
- *Theory construction lesson:* There is an extreme ambiguity as to which structure is subject to quantization upon quantizing a theory like GR.
- *Metaphysical lesson on relationism:* Quantum EPS can be understood as a relationist/structuralist project for quantum spacetime. (Related to this, we present towards the end of this chapter some reflections on how our work bears upon the hole argument of GR.)

3.1 Core features of classical EPS

To remind the reader, classical EPS is characterised by three core attributes:

- *Constructivist* in the sense of Carnap (1967, §2) (note that Carnap uses the word 'constructional' where we use 'constructivist'): the approach offers a step-by-step construction procedure based upon the principle of semantic linearity (recall again our discussion in §1.2).
- *Constructive* in the sense of Reichenbach (1969 [1924]): The basis consists of immediately empirically accessible objects or quantities. Arguably, this also implies that statements are of local nature (locality).
- *Kinematical*: The (generalized) scheme is concerned only with setting up a kinematical space for GR (spacetime theories more generally), not its dynamics.[7]

We will see that it may not be possible for a quantum version of EPS to maintain all of these principles; in particular, it may be necessary to replace constructive elements by structures less amenable to direct empirical access. However, we reiterate that the classical EPS construction also diverges from these principles at times—in particular in setting up the basic manifold structure but also at various other steps when tacit assumptions are smuggled in (see Chapter 1).

[7] See Curiel (2016) and March (forthcoming) for accounts of the kinematics/dynamics distinction, and Linnemann and Read (2021b) for some further discussion.

3.2 Point identity in quantum EPS

The central challenge for a quantum version of EPS has, we contend, to do with the notion of spacetime point identity. Specifically, once we allow superpositions of different spacetimes there is a prima facie difficulty regarding the way in which spacetime points from different branches can be mapped onto each other. And this is not merely a philosophical issue, since we apparently need to be able to map points from different branches onto each other if we are to make sensible predictions for quantum-gravitational experiments, including various recently proposed 'tabletop' experiments, which involve superpositions of branches which are subsequently recombined into a single branch—the possibility of such recombination seems to rely sensitively on the possibility of cross-branch spacetime point identifications. Indeed, we shall see that this issue arguably afflicts even the kinds of superpositions with which we are familiar in ordinary quantum mechanical experiments. Thus in this chapter we will discuss the issues involved with this question, and review a range of proposals for tackling it.

3.2.1 Point identities in classical manifolds

The starting point of the EPS construction is a set of spacetime points M (not yet structured as a differentiable manifold) which are initially postulated as distinct individuals despite the fact that at this stage there is no physical structure to distinguish them.[8] If these mathematical spacetime points are regarded as representing real physical spacetime points endowed with primitive identities, then it appears to deliver us a kinematics which would potentially be vulnerable to the hole argument (although of course in order to actually make the hole argument we also have to add dynamics to the theory)—we will return to this issue later.

However, in fact this starting assumption is arguably harmless, because once the manifold has been constructed by appeal to the behaviour of particles and light rays, we may lift the assumption of primitive identities by supposing that points are identified only by the particles and light rays that pass through them. We may then think of the actual set of events as referring to

[8] A constructivist programme *à la* EPS is restrictionist in nature: notions like 'event' are initially defined sparsely (and thus permissively), and only then does the notion get more and more restricted through the introduction of empirically motivated structure. In this restrictionist sense, the EPS programme is in the spirit of Klein's *Erlangen* approach to geometry—for a modern discussion, see Wallace (2019).

the domain of *physical* radar coordinates—and physical radar coordinates alone: each point of a physical radar coordinate neighbourhood is a (possible) crossing of light rays. In particular, a change between such physical radar coordinates can be understood as inducing only a trivial spacetime diffeomorphism, i.e. a diffeomorphism that does nothing (rather than just leaving the equations invariant).[9]

In this manner, it is reasonable to regard EPS' use of spacetime points as a convenient mathematical stepping stone. In fact, the EPS construction can therefore be regarded as an example of Cao's notion of 'self-consistent bootstrapping' (Cao, 2001), which he advocates as a solution to the problem of how we can understand the gravitational field as ontologically prior to the manifold 'spacetime', despite the fact that the latter is usually required to define local degrees of freedom to begin with (i.e. the gravitational field is formulated as a field on the manifold).[10] The idea is that we start with 'a non-physical bare manifold, on which parameter localisation can be tentatively defined', then build up our theory on that manifold, and finally argue: 'If the final results in a diffeomorphism covariant theory are independent of any specificity of parameter localisation, except for some most general features of the gravitational field [. . .] then the whole procedure is a justifiable way of investigating the physical spacetime and its ontological underpinnings' (p. 194). EPS can be read as following a similar approach, except that instead of starting from a manifold they begin with a set of bare points and derive the manifold structure there from by imposing empirically motivated axioms upon that set: since the final construction does not require one to identify spacetime points across different possible worlds, there is no longer any need for these points to have primitive identities, so the bootstrapping has succeeded.

3.2.2 Point identities in manifold superpositions

In constructing the kinematics for a theory of quantum gravity, the primitive identity question has additional complexity due to the possibility that we will get superpositions of spacetime structures, which raises the question of

[9] This is not EPS' own reading of the radar coordinates; EPS themselves employ radar coordinate charts as part of the usual manifold atlas. This means that they will be linked to non-trivial diffeomorphisms and thus, provided the dynamics is diffeomorphism-invariant, will be subject to hole argument-type objections as well.

[10] A similar issue has been discussed in the context of the dynamical approach to spacetime theories promulgated by Brown and Pooley (see Brown (2005); Brown and Pooley (2001, 2006))—see Norton (2008); Menon (2019); Chen and Fritz (2021); Linnemann and Salimkhani (2021), as well as our discussion in Chapter 2.

whether we can identify points across branches. A naïve substantivalist about spacetime would perhaps want to insist that points have primitive identities which allow for a unique identification of points across branches—even though the actual identification still has to be chosen and arguably well justified. In contrast, their structuralist counterparts who take it that spacetime points have their identities only in virtue of the behaviour of light and matter in their vicinity will presumably deny the possibility of such a straightforward identification map. Now, those falling into the latter of these camps may still attempt to employ the bootstrapping method advocated by Cao—assuming primitive identities for points at first and subsequently lifting that assumption—but it's less clear that this will succeed in the quantum context. The reason we were always free to make such a move in the classical case was that the kinematics produced by the EPS construction involves no internal modal notions, i.e. within a representational context each model represents a single possible world and its interpretation does not require reference to any other possible world, and therefore the construction will clearly never require any claims about identities of spacetime points across different possible worlds represented by one kinematical model. On the other hand, the kinematics produced by the quantum EPS construction will presumably involve distinct branches of the wavefunction each containing some spacetime points, and typically in quantum mechanics we expect that different branches of the wave function are all included in a single unified kinematics, so it's reasonable to think this construction would potentially involve identities of spacetime points across different possible branches and in this sense worlds represented by one kinematical model. So if we are not willing to accept primitive identities of spacetime points, we will have to decide how to deal with this.

Can we simply employ a model which incorporates all the different branches in a single kinematical space but refrains from postulating any identities between points in different branches? This is likely to lead to problems if we ultimately expect to end up with kinematics *directly* suitable for a theory like low-energy perturbative quantum gravity.[11] For example, it is widely expected that quantum-gravitational effects are in principle detectable in scenarios like the proposed Bose–Marletto–Vedral (BMV) experiment (Bose et al. (2017); Marletto and Vedral (2017)), which involves putting two massive particles in spatial superpositions, giving rise to four distinct branches of the wavefunction, each corresponding to a different configuration of the

[11] We discuss in the following sections the degree to which the physical setup of the dynamics actually has to be sensibly constrained by the kinematical setup.

gravitational fields sourced by the massive particles. This experiment has not yet been prepared, but most physicists expect that when it is performed we will observe entanglement forming between the two pairs of particles, mediated by the superposed gravitational fields sourced by the particles. Moreover, both in standard general relativity and in the EPS formulation, 'the gravitational field' is simply identified with the structure of spacetime; so if this experiment has the expected result, the natural interpretation in the EPS formulation is that it succeeded in creating a superposition of different spacetimes in different branches of the wavefunction. But of course it is crucial to this experiment that these distinct branches are ultimately recombined into a single branch at the end of the experiment, since otherwise we would not be able to gain access to both of the particles at once in order to measure them to establish whether or not entanglement has formed. And thus the standard prediction for the outcome of this proposed experiment assumes explicitly that it is possible to create something like a spacetime superposition and then to combine the branches back into a single branch, an assumption which follows naturally from the standard formalism of low-energy quantum gravity. Yet if there are no identities between points in different branches, it's unclear how this could possibly be done. For in standard quantum mechanics, when we create a superposition in which a particle follows two different trajectories in different branches of the superposition, the only way to 'recombine' those trajectories back into a single branch is to arrange for the trajectories to converge such that the two 'copies' of the particle subsequently have exactly the same spatiotemporal trajectory and thus are no longer distinguishable. But without identities between points in different branches, what could it possibly mean for the trajectories taken by the particle in different branches to 'converge', or for the copies to subsequently have the same spatiotemporal trajectory? On the one hand, if our approach to modelling local branching involves defining the two paths on distinct spacetimes, it seems hard to understand how the copies of the particle could possibly have the 'same' trajectory thereafter. It would seem that if branching means that particles or photons go into different spacetimes, then they should never be able to interact again—they now live on different manifolds, so there's no obvious way to define something like a scalar product between the corresponding wavefunctions of three-metrics embedded in different manifolds. And on the other hand, if our approach to modelling local branching instead puts the two paths in a single non-Hausdorff spacetime, we will still need a criterion determining when and how the branches of this spacetime join back up, and such a criterion would likely still require something similar to point identities across branches. Of course, it's possible that there exists some way of avoiding the use of point identities by instead adopting something akin to a relationalist approach—indeed, we will shortly discuss an approach of this kind based on Barbour's relationalism—but certainly at this stage in the development of the theory it remains unclear exactly how to implement such a thing.

We also need to consider the possibility of what Anandan (1997) dubs 'quantum diffeomorphisms', which are individual diffeomorphisms applied separately within different branches of the wavefunction (and the associated manifolds)—so that one may have, for example, a quantum diffeomorphism which acts as the identity in one branch, but non-trivially in another (this will be relevant in our discussions of the hole argument below).[12] Now, one important reason to think that a classical theory should be invariant under diffeomorphisms is because we don't typically suppose that classical spacetime points can be identified across different possible worlds, so performing a diffeomorphism can't possibly make any difference to the physics of the situation; and on that basis, it seems natural to think that if spacetime points can't be identified across branches, then performing a *quantum* diffeomorphism can't possibly make any difference to the physics of the situation, so we should expect that a theory of quantum gravity will be invariant under quantum diffeomorphisms. Thus this seems like an appropriate generalization of Einstein's principle of general covariance for the quantum gravitational case. But seeking to write down a theory which is invariant under quantum diffeomorphisms will likewise cause difficulties for phenomena in which systems in different branches are supposed to interact, interfere, and/or recombine—for there are good reasons to think that all of these processes can only occur locally, i.e. when the systems in question are spatiotemporally coincident, and yet an interaction between particles in different branches that is local in one choice of coordinate system will cease to be local under quantum diffeomorphisms. So if we really expect that the theory should be invariant under quantum diffeomorphisms, then prima facie it would seem that we will have to give up the demand that interactions must be local, which is a significant theoretical cost and indeed potentially incompatible with other theoretical frameworks—e.g. the locality of interactions is an important founding principle of quantum field theory. This suggests that if we don't have identities between points in different branches, we can't appeal directly to locality to understand when and where phenomena like recombination happen, which makes it a little hard to see how such a thing could possibly happen at all.

[12] We discuss further quantum diffeomorphisms and their relation to general covariance in §3.A.

Indeed, even if the BMV prediction and other predictions of low-energy perturbative quantum gravity are wrong, there are still problems, because the issue goes beyond particular scenarios involving quantum gravity phenomenology. For once we believe that some spacetime superpositions are possible, it seems natural to expect that any time a massive object is in a spatial superposition, no matter how small the mass, the spacetime in which it lives must split into two separate spacetimes—for the standard picture tells us that all masses source gravitational fields, so even if the mass is very small, nonetheless the gravitational fields and hence spacetime structure must be at least a little different in branches where it has different positions. Of course, for the small masses we deal with in current quantum experiments the difference between the spacetimes is experimentally insignificant and thus it is typically assumed that we can completely discount any effects of gravitational back-reaction. But the *magnitude* of the difference between the branches seems irrelevant to this issue of identifying points across branches, because identity is not typically regarded as coming in degrees, so it would seem that as soon as the spacetimes in the branches are even very slightly different they must now be regarded as distinct manifolds, meaning that we face all the problems described above with respect to the problem of recombining them. So if there is no way to identify points across different branches, one might naturally worry that as soon as *any* massive object enters a spacetime superposition the branches will never be able to interact again. But of course we frequently perform experiments where masses enter spatial superpositions and are then recombined—e.g. this occurs every time we put a massive particle into an interferometer where it enters a superposition of two different paths, with subsequent interference between these branches demonstrating that recombination has indeed occurred. So even based on existing experimental evidence we know that if spacetime superpositions are possible it must be possible for their branches to recombine, at least in the case where the differences between the two spacetimes are small.

Moreover, if we have total freedom to perform arbitrary quantum diffeomorphisms, then there's also the worry that in many relevant cases we will usually be able to perform separate diffeomorphisms in each branch until the spacetime structure looks the same in all the branches. For example, Anandan (1997) considers a case in which we have superpositions of different spacetime structures around a 'cosmic string' which has a superposition of different angular momenta. For each of the branches of this superposition we can find a gauge in which the metric is flat in each simply connected region outside the string, and yet a neutron approaching the string will undergo intensity oscillations as a result of this superposition, whereas it would not oscillate if it were simply in a classical region with a flat metric. So the superposition must be taken seriously as an element of reality even though it can apparently be transformed away by a quantum diffeomorphism. This demonstrates that if we want to set up a theory of quantum gravity in such a way that there is freedom to perform arbitrary quantum diffeomorphisms, these diffeomorphisms must be implemented very carefully in order to ensure that the physical predictions are preserved under diffeomorphisms, even when the differences between spacetime structure can be transformed away. In Anandan's case, the solution is to pay attention to the way the Hamiltonian constraints transform under the diffeomorphism: the same commutation relations are obtained before and after the diffeomorphism and thus the physical predictions are the same. More generally, in principle we would expect that even if a superposition state of manifolds can be brought into a form such that the metric is alike on all of them, the initial difference in metric structure will have been transferred into the matter sector: there should be no way to make metric and matter sector look the same, if the initial differences in metric structure have a substantive physical effect.[13]

These considerations suggest strongly that the theory we're building towards will ultimately need a way to identify at least some points across different branches. Of course, we don't necessarily have to implement such a thing at the level of the kinematics: as already alluded to above, it could be introduced dynamically. That the space of kinematical possibilities vastly underdetermines the range of dynamical possibilities lies in its very nature—the kinematical space is after all the arena relative to which to define the dynamics. Moreover, in the formulation of a theory there is often some freedom to shift constraints between the dynamics and the kinematics. For example, if we want our theory to be invariant under the transformation which adds 2π to some quantity θ, we can either set up our kinematics such that θ is defined on a circle and hence is periodic (kinematical symmetry), or we can set up a very general kinematics such that θ can take any value on the real line and then demand through the dynamics that all observables are periodic functions of θ (dynamical symmetry). Spekkens (2015) in particular has argued that as a consequence of this freedom of choice, the distinction between kinematics and dynamics is not empirically grounded and should give way to other paradigms; this, indeed, is also part of the moral of the 'dynamical approach' to spacetime theories (see Brown (2005)). On the other hand,

[13] This observation reinforces that the issue of linking up 'branching spacetimes' is not just about relating metric structure but rather about relating events as encoded by all field contents together.

this somewhat conventionalist approach to the kinematic/dynamic distinction may be opposed by the viewpoint of kinematical statements as constitutive (and thus, in a sense, non-empirical) statements only on top of which dynamical statements can be formulated to begin with (see Curiel (2016)). And, surely, even if there is some vagueness around the question of where to draw the line between the kinematical and the dynamical, there might still be value in using the distinction to characterize physical theorizing.

For example, one possible approach to physical theorizing involves creating an extremely general kinematics which allows for all sorts of unphysical constructions, and then imposing *any* expected symmetries and invariances at the level of the dynamics only. For example, in a sense this is essentially the approach taken in the procedure of canonical quantization, in which we first construct the most general possible Hilbert space and subsequently impose constraints to arrive at a 'physical' Hilbert space which includes only physically real possibilities (Kuchar, 1993). If we were to adopt this kind of strategy in the case of quantum EPS, constructing the most general kinematics would likely lead us to something that looks like an approach spearheaded by Hardy (2019), i.e. a bundle of manifolds with maps between them which ultimately are to be regarded as being devoid of physical significance. We would then impose invariance under quantum diffeomorphisms as a constraint on the dynamics, and also use the dynamics to model 'splitting' and 'recombining' by imposing laws with the consequence that a given pair of manifolds has identical matter content except inside some (not necessarily connected) region, so the manifolds can be regarded as one and the same except inside that region. So the kinematics would postulate many distinct manifolds, and identity between (parts of) these manifolds would be fixed by the dynamics. This approach would have the advantage that the construction of the kinematics can be done entirely locally via a construction very similar to the original EPS approach, leaving more exotic features to be encoded in the dynamics.

However, there may be reasons to prefer a different approach. For a start, we note that in classical theories the notion of 'local' is kinematical rather than dynamical—that is to say, although questions about whether the interactions allowed by the theory are local must await the specification of some dynamics, the question of what counts as 'local' in the first place is settled at the level of kinematics. So one might naturally think that in the quantum case too it should be necessary to determine what counts as 'local' for interactions between branches before defining any dynamics. Moreover, one might hope that coming up with a kinematics which more closely reflects the symmetries of the dynamics might ultimately make it easier to actually come up with the correct dynamics. For example, arguably the main point of difference between loop quantum gravity and quantum geometrodynamics/quantum general relativity is a different choice of classical kinematics—loops versus 3-metrics/4-metrics—and thus far the loop approach has concretely been more successful; so there is an argument to be made that paying attention to these sorts of issues during the construction of the kinematics may be preferable to deferring it all to the dynamics.

We also note that if one accepts that the local construction of EPS leads only to a kinematical space of superpositions of manifolds but does not itself relate points across manifold structures, then one has to show that there can be well-defined interactions for various kinds of maps between the various superposed manifold structures in some other sense. In fact, the kinematical space seems for instance still sufficient to write out regularities and find a best system relative to them without a *presupposed* standard of identifying physical structures across the manifolds. (That one can indeed find a well-defined and satisfactory best system, however, needs to be demonstrated. It might be helpful to draw on an analogy to Huggett's regularity relationalism (Huggett, 2006) and, more generally, super-Humeanism (Esfeld and Deckert, 2017).)

So, let us consider now in more specific detail the central options for dealing with the problem of identifying points across branches. In brief, we take the different options to consist in the following:

1. Prevent significant spacetime superpositions from forming at all.
2. Identify points across branches based on a criterion of similarity.
3. Identify points across branches by stipulation, potentially independently of physical goings-on.

Examples of each of these strategies have appeared in the literature, so we will now examine them in turn.

No large superpositions: Penrose's collapse approach

Penrose (1996) advocates option (1) above. He invokes the equivalence principle, which he interprets as telling us that 'it is the notion of free fall which is locally defined', and thus he contends that, if we want to map one spacetime onto another, the natural way to do that is to insist that their geodesics map onto one another, at least locally. Penrose then observes that for general superpositions of spacetime structure it will not be possible to find such a map, so the prospects for a natural standard of cross-branch identity don't seem promising. Moreover, Penrose considers the absence of such a map to be a serious problem, because it blocks the construction of a global time-translation operator, so we have no way of defining the states of well-defined energy (which are usually defined as the eignestates of the time-translation operator). Inspired by the standard theory of unstable particles, Penrose takes the absence of a well-defined energy to mean that the state is unstable, with a lifetime of $\hbar$ divided by an appropriate measure of the energy uncertainty.[14] He therefore argues that as soon as superpositions become large enough to produce interestingly different spacetime structure in each branch, they will become unstable and undergo a gravitationally induced collapse. Thus, in Penrose's scheme, ultimately we will never have to deal with the question of how to map points across branches, because as soon as the superposition becomes large enough that there is no longer a natural way to do this, it will immediately collapse.[15]

Dynamical identification: Barbour's best-matching approach

Barbour (2001), on the other hand, feels that Penrose 'is trying to solve a problem that has already been solved'. The solution, according to Barbour, is option (2) above: specifically, we can define cross-branch identities using the very same best-matching procedure that Barbour has already developed to explain how three-dimensional slices are put together into a four-dimensional object in his 'Machian relationalist' alternatives to Newtonian mechanics and general relativity (see (Mercati, 2018) for a recent book-length review). This is done by defining a measure of difference (e.g. the difference between the values of the 3-metric at matching points, integrated over all points) and seeking a set of identities between points such that the global measure of difference is extremal. The configuration which extremizes this measure of difference is the best match. Moreover, Barbour has argued that 'the three momentum constraints really do express the guts of Einsteinian dynamics and show that it arises through the creation of a metric on superspace by best-matching comparison of slightly different 3-geometries' (p. 211). Given that GR already appears to make use of best matching in this sense, Barbour argues that it makes sense to use best matching in quantum gravity as well: that is, in Barbour's view the problem of finding identities between points across different

[14] One might wonder about the extent to which a global time evolution is to be expected in a theory of quantum gravity. Arguably, Penrose's posit could, however, be read charitably as claiming that an approximate standard of global time evolution has to become available at some point towards lower energies (albeit not necessarily in the deep quantum gravity regime).

[15] Somewhat related to Penrose's proposal: according to the 'Montevideo interpretation' of quantum mechanics, quantum gravity imposes fundamental limitations on the accuracy of clocks, in turn implying a specific type of decoherence which leads to wavefunction collapse. For a philosophical appraisal of this proposal—as well as an argument to the effect that the Montevideo interpretation is currently best understood as a version of the Everett interpretation—see Butterfield (2015).

branches of the wavefunction is exactly the same as the problem of tracking 'the same point' across time in general relativity.

That said, there does seem to be a difference between these two applications of best matching. In the case of matching points across time, arguably nothing actually depends on the best matching identification of points, because all our current field theories obey a locality requirement which ensures that fields only interact when they are co-located, and hence there can be no interactions between fields at spacetime points in different time slices. Thus, although the best matched construction makes everything look simpler, in principle we could use a more complicated construction based on some arbitrary choice for how to identify points across spacetimes which would predict the same empirical results. So the door is open to interpret best matching as a conventionalist fact about the way in which we experience and construct spacetime and formulate our laws, rather than a tool which is used by nature itself. On the other hand—and as we have already seen—in the branching case the identification of points across branches would seem to have real empirical consequences if we agree that objects in different branches are supposed to interact and the branches to recombine, with the interactions and recombinations being local, i.e. taking place at 'the same points' in each branch. Assuming that we're not willing to let go of that locality assumption, it would seem that using different identities across branches would produce different predictions, since we would end up with different combinations of fields interacting. Consider for illustration the BMV experiment in which two masses in superposition states are taken to get entangled with each other through gravitational interaction (and gravitational interaction alone): the relevant literature tacitly assumes that the location of the masses are all relative to one joint lab frame—no matter whether the experiment is modelled through a Newtonian potential or (low-energy) metric fields, as done by Christodoulou and Rovelli (2019). But if we perform a quantum diffeomorphism which shifts the point at which the recombination occurs within one of the branches, then the phase change will be different and the interference effects will change. So if we use best matching to determine the point of interaction across different branches, we are no longer free to interpret it as a fact about the way in which we ourselves construct spacetime: nature itself must make use of best matching.

Moreover, Barbour's best matching approach looks difficult to implement within the EPS scheme for two reasons: (i) Barbour's best matching requires us to compare the whole of a three-geometry all at once, and therefore we don't have access to that kind of procedure if we are purely dealing with local kinematics. And (ii) best matching requires us to employ the actual metric, so it most likely has to be implemented at the level of the dynamics. So if we accept Barbour's approach to cross-branch identities, it would seem that we have to give up on the ideas that 'locality' can be defined at the level of kinematics and that it is a meaningful concept to the local observer herself!

A possible approach to option (2), i.e. dynamical identification, other than that of best matching would be to invoke an effective field theory description. Here, we split the metric into a constant background $\eta_{\mu\nu}$ plus some small fluctuations $h_{\mu\nu}$, where $\eta_{\mu\nu}$ is constant across all the branches so only the fluctuations $h_{\mu\nu}$ are quantized. We can then use $\eta_{\mu\nu}$ to define a 'background' which tells us where the particles recombine. Several options are then available. We could adopt a description similar to that suggested by Penrose, where spacetime superpositions can occur provided that the difference between the spacetimes is small enough that the EFT description is possible; but once the differences become too large for this description to work, the superposition must collapse. Alternatively, we could say that spacetime superpositions can only be recombined as long as the differences are small enough for the EFT description to be valid; once the branches become too different, it's no longer possible for them to be recombined, so we get a decoherence-like effect and the branches become separate non-interacting Everettian worlds. However, both of these options seem a little troubling, because even when the difference between the spacetimes is small, the identification between the spacetimes given by $\eta_{\mu\nu}$ is only an approximate matter, and yet it would seem that the resulting identity map which determines where the recombination occurs can't be an approximate matter. Also, it seems odd that the possibility of recombination should depend upon how similar the branches are, because recombination itself doesn't seem to come in degrees: either the branches come back together into one spacetime or they split and subsequently never interact, so there isn't really any intermediate option. Thus the EFT route forces us to postulate a very sharp transition between 'sufficiently similar' and 'not sufficiently similar', even though similarity between branches is something that presumably varies along some sort of continuous scale. Moreover, these options would seem to compel us to espouse either a collapse model or the Everett interpretation, so the EFT approach may not be appealing if we want to avoid collapses and we also want to avoid multiple worlds.

Kinematical indifference: Hardy's q-diffeomorphism approach

Hardy (2019) takes up option (3) in the context of his work on quantum coordinate systems: he advocates a 'gauge description', by which is meant that we arbitrarily choose some way of mapping points from one branch to another.

This enables him to arrive at a scheme of 'quantum manifolds', in which each branch of the wavefunction corresponds to a distinct manifold. (Note that Hardy does not prove that the quantum manifolds are in fact manifolds in the usual technical sense—but potentially the EPS scheme could be used to do that.) The coordinate systems on these manifolds define a mapping between them, but the mapping is purely conventional and nothing should ultimately depend on the choice of map. It follows that the final theory should indeed be invariant with respect to quantum diffeomorphisms, which (recall) are diffeomorphisms whose action can be different within each branch of the wave function.[16]

This means that if we do take up option (3), we will have to insist that all solutions be invariant under quantum diffeomorphisms (let's call these 'q-diffeomorphisms' in what follows). In particular, if identities between spacetime points in different branches are indeed to be regarded as unphysical—as a mere 'gauge choice'—one might wonder whether our standard techniques for dealing with systems exhibiting gauge freedom can be extended to this context. Most gauge theories of interest are associated with constrained Hamiltonian systems, where the presymplectic phase space N of the physical theory arises as a regular submanifold of a symplectic geometry. In the case of canonical general relativity, N is defined by a set of four constraint functions per space point; the three momentum or vector constraints, which enforce diffeomorphism invariance, and the Hamiltonian or scalar constraint, which enforces invariance with respect to global time translations. When we quantize the theory, in order to maintain gauge invariance, we turn each first-class classical constraint into a quantum operator C and insist that the space of physical states must obey the constraint $C|\psi\rangle = 0$—i.e. the constraints still vanish on the physical Hilbert space. Let us now take as an example the geometrodynamical approach to quantum gravity which treats the 3-metric $h_{\mu\nu}(x)$ as a field defined on a manifold and then quantizes it to give $\hat{h}_{\mu\nu}(x)$. The field eigenstates $|e\rangle$ are the states $\hat{h}_{\mu\nu}(x)|e\rangle = h_{\mu\nu}(x)|e\rangle$, so each eigenstate describes a configuration across spacetime. We can define a wavefunctional $\Psi[e] = \langle e|\Psi\rangle$, i.e. the wavefunctional gives the amplitude for any field configuration, which is to say that in general it describes a superposition of field configurations. Due to the diffeomorphism constraint, $\Psi[e]$ is invariant under coordinate transformations in 3-space, i.e.

[16] Anandan (1997) makes roughly the same point, suggesting that invariance under quantum diffeomorphisms can be regarded as a quantum principle of general covariance. Giacomini et al. (2019) discuss a similar issue within the 'quantum reference frame' research programme, although their quantum reference frames are defined at a given time, whereas Hardy works with coordinate systems for entire spacetimes.

the same amplitude is assigned to a configuration and to its shifted version. Thus, as long as different branches in the superposition of spacetime structure are completely independent and non-interacting, the wavefunctionals and the theory more broadly are invariant under the action of q-diffeomorphisms.

However, it's unclear that the branches are in fact non-interacting: as noted above, typically it is expected that we should be able to see interference effects between different branches of the wavefunctional. Because interference between distinct branches is not possible within the classical theory, the classical constraints aren't designed to enforce invariance of interactions between branches under q-diffeomorphisms, so it is not obviously the case that the quantized versions of these constraints will successfully enforce invariance with respect to q-diffeomorphisms. Indeed, it can be shown that Dirac quantization, in which we first quantize the theory and then apply the constraints, is equivalent to reduced quantization,[17] in which we first apply the constraints and then quantize the theory, at least in the one-loop approximation. Clearly, if we use reduced quantization the constraints cannot yield covariance with respect to quantum diffeomorphisms, because the structure that makes quantum diffeomorphisms possible was not available prior to quantization. Thus it seems likely that Dirac quantization also will not yield invariance with respect to quantum diffeomorphisms, so there remains a worry that the standard constraint quantization used in the construction of quantum theories of gravity will not in fact always remove all the gauge degrees of freedom.

Further difficulties arise as soon as we include interactions with other fields. All the important field theories of modern physics are local, i.e. they postulate only interactions of the form $\phi(x)\chi(x)$ defined between two fields at the same point. So in the quantum field theory we'd expect to see coupling terms of the form $h_{\mu\nu}(x)\phi(x)$ where $\phi(x)$ is some matter field coupled to the metric. Again, this will work so long as we're happy for $\phi(x)$ to split up into distinct branches with each branch coupled only to the metric inside its individual branch and no interactions between different branches; but in reality we expect the matter field $\phi(x)$ to exhibit interference between its branches, and since interference is a local phenomenon it would seem that performing diffeomorphisms inside one branch and not another might change the resulting interference of the matter field $\phi(x)$. In order to avoid this, we would presumably need to require that the q-diffeomorphisms go smoothly to zero at all points where matter fields exhibit self-interactions, and yet we can't enforce

[17] See nLab authors (2022).

that requirement without first having a standard of cross-branch identity of points, since otherwise we will have no idea which fields $\phi(x), \phi(x')$ we are supposed to match up.

One possible way to write down a kinematics for quantum gravity such that it will indeed have its simplest expression in a form which is covariant with respect to q-diffeomorphisms may be to write it in terms of closed loops, or holonomies, since closed loops are preserved by diffeomorphisms.[18] This of course is exactly what loop quantum gravity does. Because the canonical variables of LQG are loops rather than metric fields at points, no physical content is encoded in the values of the metric at specific spacetime points, and therefore we will never have any need to identify the same point across different branches.[19] However, this looks like a difficult option to implement within an EPS-style scheme: the EPS scheme is originally intended as an approach to reducing *local* kinematics to empirical observations on light and (free) particle movement—a kinematics that is not local is thus by definition out of its reach.

3.3 Variants of quantum EPS

Following the discussion in previous sections, we now propose several different ways in which one might approach a quantum EPS construction. None of these options is necessarily right or wrong: there is simply a choice to be made about how much structure we want to put into the kinematics versus the dynamics, and about what exactly we mean by 'spacetime' in the first place. Importantly, all these proposals will have to circumvent the obstacles regarding the identities of spacetime points which were raised in the previous section. In fact, each variant we present in the following is inspired by one of those proposals (1–3) for dealing with these obstacles. More precisely, in §3.3.1 we suggest a strategy which proceeds by ruling out the existence of superpositions of spacetimes (like Penrose's approach); in §3.3.2 we discuss a strategy based on non-local features (like Barbour's approach); and in §3.3.3 we suggest an approach which defers most of these matters to the dynamics (like Hardy's approach).

[18] The move by Westman and Sonego (2008) to construct scalars and use them to span a new, diffeomorphism-invariant 'beable' space, and Earman's 1977 notion of Einstein/Leibniz algebras, might be suitable (Kretschman objection-resistant) generally covariant alternatives to the metric as well. Notably, in particular for the Einstein/Leibniz algebra approach it is still controversial, though, as to whether this approach does not just suffer from an analogous issue with diffeomorphism invariance.

[19] See for instance Rovelli (1991), in particular §1.5, for a definition of the loop variables.

3.3.1 Option (i): Operationalism about spacetime

One of the lessons of the foregoing discussion is that gaining operational access to different branches of a spacetime superposition is extremely non-trivial. As a result, one might start to question whether it makes sense to expect the operationally motivated axioms of EPS to obtain within each individual branch of the spacetime, given that not all of these branches can have operational significance.

Motivated by this, another possible way of thinking about quantum gravity kinematics would be really to commit to the operational approach and argue that it has the consequence that we can't have 'spacetime superpositions' at all. That is, spacetime is to be understood as an emergent structure defined in terms of what is directly operationally accessible to us,[20] and thus since superpositions of spacetimes can be detected only indirectly (e.g. via measurements of entanglement as in the BMV experiment), they do not in any concrete sense involve 'real' spacetimes.

This emphasizes the fact that there are two different possible ways of thinking about the emergence of spacetime within quantum gravity. Many current approaches to quantum gravity agree that spacetime as we experience it should be understood as arising from an underlying substratum of quantum 'stuff' which is not defined on any spacetime. There are, however, two possible routes that this could take: we could imagine that first a 'quantum spacetime' (which can participate in superpositions) arises out of the substratum, and then our ordinary single-valued spacetime emerges from the quantum spacetime in the macroscopic limit, or we could imagine that our ordinary single-valued spacetime arises directly out of the substratum, with—at least for all effective purposes—no intermediate layer of 'quantum spacetime'. The former approach is presupposed in the standard analysis of low-energy quantum gravity tests like the BMV experiment where we are invited to suppose that two different spacetimes are superposed. The latter approach would have the interesting consequence that although gravity is due to quantum structure, we nonetheless never get superpositions of different spacetimes. This approach has its attractions: in particular, it would sidestep the whole question of identity of spacetime points across branches, since 'spacetime points' would only be defined in the macroscopic limit where branches become non-interacting due to decoherence.[21] Indeed, this approach would lead to the conclusion that

[20] Or, at least characterized to some extent in terms of what is operationally accessible.

[21] Such an approach is realized in various emergent gravity approaches. For a critical survey, see Linnemann and Visser (2018).

the decoherence mechanism is an essential feature of the emergence of spacetime, and thus in regimes without decoherence we shouldn't expect to arrive at manifold structures of the kind constructed by EPS.

The strict operational approach has precedents in previous work on operational features of general relativity. In particular, we recall that the connection between the operational features probed by EPS and the metric field is typically understood in terms of (a version of) the equivalence principle: in Knox's words (Knox, 2013), the equivalence principle 'expresses just that fact about our matter theories that must be true if systems formed from appropriate matter are to reflect the structure of the metric field, that is, if phenomenological geometry is to reflect the geometry of the metric field' (p. 350).[22] Knox thus argues that since the equivalence principle can only be defined in an approximate and contextual way, it follows that phenomenological geometry is coarse-grained, because 'operationalized reference frames are objects of finite spatial extent and therefore can't perfectly instantiate metric structure' (p. 352). Thus any spacetime defined in terms of phenomenological, operationally accessible geometry will also be coarse-grained, meaning that we will not be able to think of such a spacetime 'as a background manifold with exact geometrical properties'. If we accept Knox's argument, then it follows that the operational axioms used by EPS should not be regarded as exactly true even in the classical case: thus in both the classical case and the quantum case the axioms can be understood as characterizing high-level emergent features, which means that superpositions of spacetimes need not ever appear.

Of course, if we define spacetime in this operational way we still need some mathematical way of describing what goes on in scenarios like the BMV experiment, but in principle this can always be done using an effective field theory description. That is, we define a 'background spacetime' using directly observable operational procedures: since we never observe superpositions directly, this background spacetime is necessarily classical and single-valued. Then small fluctuations in the metric can be modelled as a quantum field defined on top of this classical, single-valued spacetime: thus the fluctuations can be in superpositions but the background spacetime itself cannot. This is how the BMV experiment is currently modelled—it is taken for granted that we have a background laboratory frame which is used to define meaningful mappings between the branches of the 'spacetime superposition'.

[22] For further discussion on the coincidence of (a) what is measured operationally via material fields and (b) geometrical structure, see Read et al. (2018).

That said, the really difficult regimes for quantum gravity are those in which no effective field theory description is possible. Taking a hard-line operational approach to spacetime suggests that, in fact, an effective field theory description is always possible, because the operationally defined spacetime must always exist and be single-valued. Problems might arise, however, when we get into regimes where the behaviour of the background spacetime can't be correctly modelled without taking into account quantum gravity effects.

What would this approach mean for a quantum EPS construction? Straightforwardly, it would imply that the quantum EPS construction would just reduce to the classical EPS construction, because all the same classical axioms will be true of spacetime at the operationally accessible level, and we would not expect the underlying non-operational structure to be captured by an EPS-style axiomatization.

3.3.2 Option (ii): non-local scenario

A second possible approach to a quantum version of EPS involves giving up the ambition of a *local* constructive approach to quantum gravity, and instead building up our spacetime from non-local objects like loops, or other objects which are guaranteed to be preserved under (q-)diffeomorphisms. Indeed, this is exactly the approach that has been taken with some success in the loop quantum gravity (LQG) programme (see e.g. Rovelli (2004)); notably, even the gravitational path integral can be understood in this way (cf. Rovelli and Vidotto (2015)).

Now, the non-local objects used in an approach like LQG are not directly suitable for use in an EPS-style construction, because they have no transparent operational interpretation. For example, although the 'edges' making up the loops in the LQG graph can sometimes be interpreted loosely as encoding some kind of proximity relation, this is not always feasible—as Rovelli and Vidotto (2024, p. 4257) explain, 'the notion of adjacency defined by the graph may not match the one implicitly defined by an averaged large geometry [. . .] there may be two distinct notions of contiguity in the theory: the one defined by the graph (that underpins the dynamics of the spin networks) and the one defined by the emergent smooth geometry.' Abstract relations of this kind would clearly not be well-suited as the basic object of a constructivist project. However, one could imagine a more constructivist version of the view based on loop-like objects defined in operational terms. Indeed, a non-local construction can be achieved quite naturally within the structures postulated by classical EPS. For we may simply include in our axioms a postulate ensuring that superpositions always or at least sometimes come to an end (i.e. the different copies of a light ray in different branches of the superposition ultimately meet and are combined back into a single light ray). We do not have to commit here to an account of how this comes about—that can be added at the dynamical stage, e.g. by best matching. Thus pairs of manifolds should be defined such that they coincide at all times except inside one or more causal diamonds, where a causal diamond is the intersection of the causal future of some spacetime point p_1 and the causal past of some other spacetime point p_2 (e.g. p_1 may be identified as the point at which some mass first enters a spatial superposition and p_2 may be the point at which the spatial branches of the mass are recombined).

This route fits very nicely into the EPS approach, because an 'echo' as employed by EPS is exactly the right kind of structure to define a causal diamond. An 'echo' in the EPS construction is a map from a point P on the trajectory of a particle to a later point Q on the same trajectory, understood to be defined by sending a light ray away from the first point, and then receiving it back at the later point after its reflection from some object. The path taken by the light ray in this case defines part of the boundary of the causal diamond defined by P and Q. e play a significant role in the EPS construction, particularly in the development of radar coordinates, and thus in effect the structure we need to implement a non-local quantum EPS is already inherent in classical EPS. Motivated by this construction, one could imagine formulating an EPS scheme which takes as its basic object not 'messages' but 'causal diamonds', which ensures that spacetime superpositions are 'anchored' in a time-symmetric way such that the resulting manifolds coincide both in the past and in the future. If we take this approach then we will not be postulating a set of distinct manifolds as in Hardy's picture: instead we will have a single manifold which branches at some points and then recombines, i.e. a non-Hausdorff manifold.

Of course, clearly this option would to some extent still involve a departure from classical EPS core principle of constructiveness, insofar as that is understood as involving a commitment to locality. Thus acceptance of this option could be interpreted as an acknowledgement that 'locality' in the usual sense simply can't be preserved within quantum gravity. However, recall that we noted in the introduction that one can simply read EPS as an ontological-relationist programme which builds up spacetime from what count naïvely as high-level notions. On this construal, not all 'high-level' notions need be immediately accessible to a single observer at a single time, although they should still be operationally definable in some more general sense; so one could still accept a 'quantum EPS' based on causal diamonds as a legitimate extrapolation of the original methodology to the quantum domain, since causal diamonds are still reasonably operational as compared to the abstract loop structures used in LQG, and thus such an approach would plausibly offer a sensible account of the operational significance of quantum-gravitational phenomena.

Another interesting possibility on the subject of non-locality would be to explore an approach based on the conjecture that spacetime structure can be derived from entanglement, which has recently been popularized within quantum foundations.[23] These proposals are based on the fact that entanglement in certain sorts of systems obeys an 'area law', which inspires the conjecture that perhaps we can arrive at a spacetime metric by writing the distance between spacetime regions as a function of the degree of entanglement between them. This approach shares with EPS the motivation of constructing spacetime out of simple first principles, but since entanglement is a paradigmatic example of non-locality, this approach will evidently not satisfy the locality criterion. But, if we take route (ii) and embrace a non-local EPS approach, one might hope to find a fruitful unification of the entanglement-based approach and EPS. One important limitation of the entanglement-based approach is that since it defines distances between regions in terms of the total amount of entanglement between those regions, it is quite coarse-grained and thus may not be well-equipped to assign different distances within different branches of the wavefunction—and this may lead to problems in cases like the BMV experiment, since it is very important to the interpretation of this experiment that the particles are indeed at different distances in different branches. But since the EPS construction has been designed to allow us to construct the metric in full detail, it is possible that supplementing the entanglement approach with some EPS-like axioms would give it the capability to deal with these sorts of cases, so a combination of the two constructions could be of great interest even if not precisely in the spirit of the original EPS approach.

3.3.3 Option (iii): laissez-faire scenario

This option involves simply deferring all questions about maps between branches and recombination to the dynamics. It is then very straightforward to

[23] See e.g. Van Raamsdonk (2010), Jaksland (2021), Cao et al. (2017).

arrive at a quantum EPS construction: essentially, we replace classical events with quantum events (a notion which we will make precise below) and then simply require that all the same axioms hold individually within each branch of the wavefunction. This setup is minimal, but at least prima facie it would seem to provide a sufficiently well-defined arena relative to which dynamics can be set up.

Although this is the most straightforward option, some doubts remain. First, one might worry that the resulting kinematics will be too general to be useful—we might well be better off with a more restricted kinematics which better reflects the constraints. Second, one might also wonder if it is really natural to expect all the classical axioms to hold individually within each branch of the wavefunction—after all, it is not straightforward to operationally access distinct branches of the wavefunction, so even though the axioms look operational we cannot actually verify operationally that they hold in the same way in other branches of the wavefunction. In particular, as we noted in the introduction to this chapter, although inductive generalization may support the idea that the axioms can be expected to hold in the case of large-scale superpositions of spacetimes which have undergone decoherence, it's unclear that they will still hold in the same way for small and mid-sized superpositions. So there are certainly open questions around the appropriateness of a laissez-faire approach, particularly if this is expected to describe superpositions too small to contain realistic observers; but nonetheless, it is educational to consider the laissez-faire approach at least as a starting point, so in the next section we will explore further what that might look like.

3.4 Explicit laissez-faire quantum EPS constructions

In this section, we present two possible starting points for a quantum EPS axiomatization based on the laissez-faire approach presented in §3.3.3. In the first, we explicitly quantize rays and particles using a Hilbert space built out of events; in the second, we take inspiration from the consistent- and decoherent-history approaches to quantum mechanics (see Griffiths (2019) for an overview) as well as the (not fully unrelated) branching spacetime literature (see Luc (2020); Luc and Placek (2020) for recent introductions). In each case we present the first few classical EPS axioms and alongside offer the corresponding axioms for a quantum EPS construction.

3.4.1 In terms of superpositions

We begin with a laissez-faire approach based upon quantized rays and particles.

1. *A point set $M = \{p, q\ldots\}$ is a set of events.*

   Q: A point set $M = \{p, q\ldots\}$ is a set of events, where each event is identified with an element in a quantum basis $\{|v_p\rangle, |v_q\rangle\ldots\}$, so M defines a Hilbert space $\mathbb{M}$ whose dimension is equal to $|M|$.[24]

2. *Light rays and particles are subsets of M.*

   Q: Light rays and particles are sets of superpositions of events, e.g. $P = \{\sum_i c_i|v_i\rangle, \sum_i c_i'|v_i\rangle\ldots\}$ where $\sum_i |c_i|^2 = 1$ where $\sum_i |c_i'|^2 = 1$. With each particle P we can associate a Hilbert space $\mathbb{P}$ which is equal to the subspace of $\mathbb{M}$ defined by all and only those events $|e\rangle$ such that some element of P has support on $|e\rangle$. Likewise, with each light ray we can associate a Hilbert space $\mathbb{L}$ defined similarly.

3. *The map $e_Q : P \to P, p \mapsto e_Q(p) =: p'$ is called an echo on P from Q.*

   Q: The map $e_Q : \mathbb{P} \to \mathbb{P}, \sum_i c_i|v_i\rangle \mapsto e_Q(\sum_i c_i|v_i\rangle)$ is called an *echo* on P from Q. The echo map is a map on just the coefficients $\{c_i\}$, i.e. it transforms only coefficients and not events themselves.

4. *The map $m : P \to Q, p \mapsto m(p)$ is called a message from P to Q.*

   Q: The map $m : \mathbb{P} \to \mathbb{Q}, \sum_i c_i|v_i\rangle \mapsto m(\sum_i c_i|v_i\rangle)$ is called a *message* from P to Q.

5. *Axiom D_1: every particle is a smooth, one-dimensional manifold and any echo on P from Q is smooth and smoothly invertible.*

   QA1: Every particle can be written as a superposition of smooth, one-dimensional manifolds, i.e. we can write $P = \sum_{c_i} c_i|V_i\rangle$ where each V_i is a set of basis elements which correspond to points that belong to a smooth, one-dimensional manifold. An echo on P from Q can be split into a superposition of *e* which are smooth and smoothly invertible, i.e. when an echo map is applied to a point $\sum_i c_i|v_i\rangle$, the result is $\sum_i c_i|q_i\rangle$ where we can identify a smooth and smoothly invertible echo from

[24] It's possible that for realistic spacetimes this basis would have to be uncountable, so $\mathbb{M}$ would be a non-separable Hilbert space, which leads to a variety of technical complexities—see Earman (2020) for details. However, in this construction we will make the simplifying assumption that the relevant basis is countable so the Hilbert space can be separable.

c_i to q_i for all i. In the case where the echo has a fixed starting point this means we end up with several different smooth manifolds with the same starting point but different middle and endpoints.

6. *Axiom D_2: any message from a particle P to another particle Q is smooth.*

   QA2: We require that any message from a particle P to another particle Q can be split into a superposition of messages which are smooth.

7. *Axiom D_3: there exists a collection of triplets (U, P, P') where $U \subset M$ and $P, P' \in \mathcal{P}$ [with $\mathcal{P}$ the set of particles] such that the system of maps $\{x_{PP'|U}\}$ is a smooth atlas for M. Each map is written in terms of coordinates (u, v, u', v') where u and v are emission and arrival times at P and likewise on P'.*

   QA3: There exists a collection of triplets (U, P, P') where $\mathbb{U} \subset \mathbb{M}$ and $\mathbb{P}, \mathbb{P}'$ such that the system of maps $\{x_{PP'|U}\}$ is a smooth atlas for M. Each map is written in terms of coordinates (u, v, u', v') where u and v are emission and arrival times at P and likewise on P'.

8. *Claim: Every particle is a smooth curve inM.*

   QC: The proof of this claim can be rewritten using the axioms above: it follows straightforwardly from linearity that every particle is a superposition of smooth curves in M.

9. *Axiom D_4: Every light ray is a smooth curve in M.*

   QA4: Every light ray is a superposition of smooth curves in M.

10. *Axiom L_1: any event e has a neighbourhood V such that each event p in V can be connected within V to a particle by at most two light rays. Given such a neighbourhood and a particle P through e, there is another neighbourhood $U \subset V$ such that any event p in U can be connected with P within V by precisely two light rays and these intersect P in two distinct events e_1, e_2. If t is a coordinate function on $P \cap V$ with $t(e) = 0$, then $g = -t(e_1)t(e_2)$ is a function of class C^2 on U (i.e. it is twice differentiable on U).*

    QL1: Because in our construction each event occurs only in a single branch of the wavefunction, for each event we can define a neighbourhood V of points that are within the same branch, and such that each event $p \in V$ can be connected within V to a particle by at most two light rays. Given such a neighbourhood and a particle P through e, there is another neighbourhood $U \subset V$ such that any event $p \in U$ can be connected with P within V by precisely two light rays and these intersect P in two distinct events e_1, e_2 (and since U is included in V it also belongs to the same branch of the wavefunction). If t is a coordinate function on $P \cap V$ with $t(e) = 0$, then $g = -t(e_1)t(e_2)$ is a function of class C^2 on U (i.e. it is twice differentiable on U).

11. *Axiom L_2: the set L_e of light-directions at an arbitrary event e separates the projective space at e into two connected components. In the tangent space at e, the set of all non-vanishing vectors that are tangent to light rays consists of two connected components.*

    QL2: The set L_e of light-directions at an arbitrary event e separates the projective space at e into two connected components. In the tangent space at e, the set of all non-vanishing vectors that are tangent to light rays consists of two connected components. (Note that all of this will take place within the same branch of the wavefunction: no relationships between the projective spaces for different branches can be postulated in the laissez-faire approach.)

12. Now we explore the properties of g and extract a rank-two tensor from it.
    (a) $\partial_\mu g = 0$.
    (b) $g_{\mu\nu} = \partial_\mu \partial_\nu g$ defines a tensor at e.
    (c) The tangent vector T_μ of any light ray L through the point e satisfies $g_{\mu\nu} T_\mu T_\nu = 0$.
    (d) $g_{\mu\nu} \neq 0$ on particle rays.

    Q: Since we have defined g using only points which lie in the same branch of the wavefunction as the original point e, this part of the construction is exactly the same as the classical case and thus we obtain a rank-two tensor defined on a local 'patch' all within a single branch of the wavefunction.

We will not go through the remaining EPS axioms explicitly here, because now that we have set up the superposition of manifolds and their coordinatizations, the remaining axioms do not have to be altered from the classical case (as shown for example with the construction of g in step 12 above).

3.4.2 In terms of branching spacetime

We turn now to another possible laissez-faire approach, this time based upon the notion of branching spacetimes.[25] The definitions of the set of events,

[25] For a recent book-length introduction to branching spacetimes and their philosophical significance, see Belnap et al. (2022).

particles, and light rays, as well as of messages and e, are similar to their counterparts in the classical EPS approach, the only difference being that the maps corresponding to messages and e in the quantum case have to be allowed to be multi-valued.

1. A point set $M = \{p, q...\}$ is a set of events.
2. Light rays and particles are subsets of M.
3. The (possibly multi-valued) map $e_Q : P \to P,\ p \to e_Q(p) =: p'$ is called an *echo* on P from Q.
4. The (possibly multi-valued) map $m : P \to Q,\ p \to m(p)$ is called a *message* from P to Q.

We then add a new axiom on branching structure:

5. **QA0:** Each particle (set) P can be decomposed as $P = \bigcup_{i=0}^{n} P_i$ with $P_i = P_{i-1} \cup \bigcup_{j=1}^{m_i} p_j^i$ and P_0, p_j^i such that messages from other particles onto P_0, p_j^i are always single-valued. The subsets of P, $B_{j_1,...j_n} := P_0 \cup \bigcup_{i=1}^{n} p_j^i$ are called *particle branches* of P.

The standard axioms need to be understood without any assumption that the resulting manifold be Hausdorff (as assumed tacitly in the original EPS construction: see Chapter 1):

6. *Axiom D_1: every particle is a smooth, one-dimensional manifold and any echo on P from Q is smooth and smoothly invertible.*

   QA1: Every particle is a one-dimensional non-Hausdorff manifold. Any echo on a branch of P, denoted by P_B, from a branch of Q, denoted by B_Q, is smooth and smoothly invertible.

7. *Axiom D_2: any message from a particle P to another particle Q is smooth.*

   QA2: Any message from a particle branch B_P to another particle branch B_Q is smooth.

8. *Axiom D_3: there exists a collection of triplets (U, P, P') where $U \subset M$ and $P, P' \in \mathcal{P}$ [with $\mathcal{P}$ the set of particles] such that the system of maps $\{x_{PP'|U}\}$ is a smooth atlas for M. Each map is written in terms of coordinates (u, v, u', v') where u and v are emission and arrival times at P and likewise on P'.*

   QA3: There exists a collection of triplets $(U, B_P, B_{P'})$ where $U \subset M$ and $B_P, B_{P'}$ are particle branches relative to P and P' such that the system of maps $\{x_{B_P B_{P'}}|U\}$ is a smooth atlas for M. Each map is written in terms of coordinates (u, v, u', v') where u and v are emission and arrival times at B_P and likewise on $B_{P'}$.

9. *Claim: Every particle is a smooth curve in M.*

   QC4: Every particle branch $B_{(j_1,\dots,j_n)}$ is a smooth curve in M with $C : [0, 1] \to M$.

   Proof: See Chapter 1 for the proof in the case of M being Hausdorff; the proof, however, does not depend on the Hausdorff assumption, and thus carries over.

10. The nature of particles can be characterized further as follows:

    QA4: Consider a particle P: it is the union of all its branches, i.e. $P = \bigcup_{(j_1,\dots,j_n)} B_{(j_1,\dots,j_n)}$; each branch $B_{(j_1,\dots,j_n)}$ is denoted by the branch index $(j_1, \dots, j_n)$ and has a corresponding curve $C_{(j_1,\dots,j_n)} : [0, 1) \to M$. Then for each such curve $C_{(j_1,\dots,j_n)}$, there exists:

    (i) $g \in (0, 1)$, and
    (ii) a branch index $(j_1, \dots j_{i-1}, j'_i, \dots, j'_n)$ such that $C_{(j_1,\dots,j_n)} = C_{(j_1,\dots j_{i-1},j'_i,\dots,j'_n)}$ on $[0, g)$ and $C_{(j_1,\dots,j_n)} \neq C_{(j_1,\dots j_{i-1},j'_i,\dots,j'_n)}$ on $[g, 1]$.

In other words, each particle P is a multifurcate curve of the second kind in M.[26]

Remarks: (1) The axiom is not genuinely empirical as only single branches of the branching structure can be subject to observation: this point has already been discussed in the previous sections of this chapter. (2) Importantly, and unlike the original EPS for axiom D_3, we do not assume implicitly that the smooth manifold established by **QA3** be Hausdorff. (Note that the manifold is therefore not guaranteed to be determined uniquely.) Given that standard differential geometry works with the Hausdorff assumption as a tacit presupposition, it is worth stressing that smooth manifolds can very well be non-Hausdorff. For a first intuitive grasp, consider how Figures 3.1 and 3.2 illustrate how charts can still be straightforwardly defined on branching lines and surfaces, providing the right intuition as to why this is so on branching (and thus non-Hausdorff) manifolds more generally as well. In particular,

[26] The branching spacetime literature standardly features *bifurcate* curves of the second kind:

> A bifurcate curve of the second kind is a pair of curves C, C' in a manifold W, with $C, C' : [0, 1] \to W$, such that $C = C'$ on $[0, g)$ and $C \neq C'$ on $[g, 1]$ for some $g \in (0, 1]$. (Luc, 2020, definition 5)

The multifurcate curve is its natural generalization.

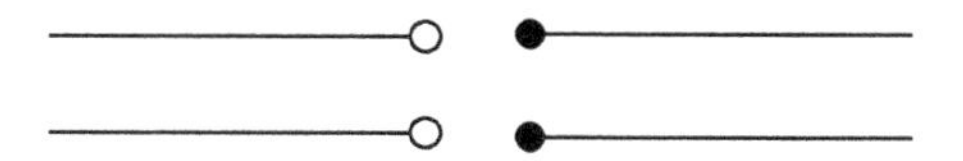

Figure 3.1 One-dimensional branching structure arising from identifying the two lines on the left-hand side

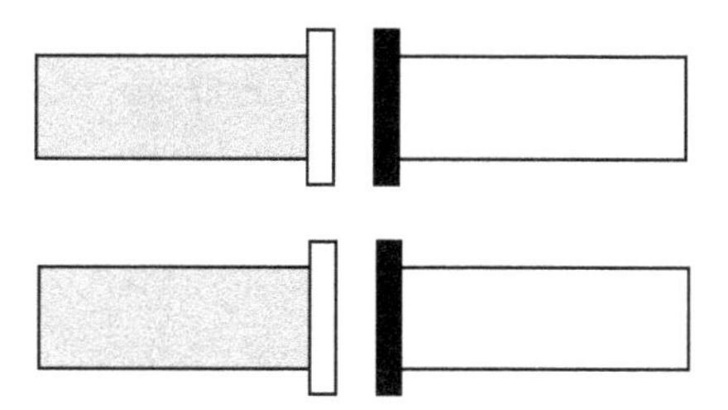

Figure 3.2 Two-dimensional branching structure arising from identifying the two areas on the left-hand side

one sees from the comparison of the surface to the line case that the idea of a submanifold—effectively a down-projection of a manifold into a lower dimension space, when seen from the correct chart—does not depend on the Hausdorff condition. (3) An alternative to (the above formulation of) **QA4** is this: Every particle segment p^i_j is a half-open curve $C : [0, 1) \rightarrow p^i_j$ with $\lim_{a\rightarrow 1} C(a) \in p_{j_{i+1}}$ for some j.

QC1: P is a submanifold of M.

Proof: This is analogous to part 1 in the proof of the analogous classical claim found in Chapter 1.

QC2: P is a non-Hausdorff submanifold of M.

Proof: This is an immediate consequence of **QA4.**

QC3: M is a non-Hausdorff manifold.

Proof: This follows from the contraposition of that submanifolds of Hausdorff manifold are Hausdorff again, and the preceding claims.

Remark: Importantly, a non-Hausdorff differentiable manifold is not uniquely determined by providing atlas structure—after all, it can have both local and global branching. However, the local branching is to be determined through observations of the observer (see below) while the global branching is the kind of fact one can be happy to stay ignorant about in a local approach (akin to how we cannot expect to learn about the global nature of the metric in general relativity).

11. *Axiom D_4: Every light ray is a smooth curve in M.*

    QA4: Every light ray is a superposition of smooth curves in M.

12. *Axiom L_1: any event e has a neighbourhood V such that each event p in V can be connected within V to a particle by at most two light rays. Given such a neighbourhood and a particle P through e, there is another neighbourhood $U \subset V$ such that any event p in U can be connected with P within V by precisely two light rays and these intersect P in two distinct events e_1, e_2. If t is a coordinate function on $P \cap V$ with $t(e) = 0$, then $g = -t(e_1)t(e_2)$ is a function of class C^2 on U (i.e. it is twice differentiable on U).*

    QL1: Any event e has a neighbourhood V such that each event p in V can be connected within V to a particle branch by at most two light rays. Given such a neighbourhood and a particle branch B_P through e, there is another neighbourhood $U \subset V$ such that any event $p \in U$ can be connected with B_P within V by precisely two light rays and these intersect B_P in two distinct events e_1, e_2. If t is a coordinate function on $B_P \cap V$ with $t(e) = 0$, then $g = -t(e_1)t(e_2)$ is a function of class C^2 on U (i.e. it is twice differentiable on U).

    Remark: One might wonder why messages/echos introduced before are not taken to lead to branching structures. Now, we can assume that light rays *qua* messages/echos are determinate, i.e. not branching themselves, given that the axiom that there is at most one signal going back and forth (L_1) for a certain neighbourhood of the particle is arguably still valid even if several signals become possible. For each signal can only differ by direction but—being lightlike—not by speed; and given that the signal has to hit a specific other particle branch from a given particle branch, a restriction from the classical case such as that at most only one echo exists seems to be unchanged by whether or not *several* light signals are actually emitted in different directions (as in the quantum case)—or just hypothetically (as in the classical case).

13. *Axiom L_2: the set L_e of light-directions at an arbitrary event e separates the projective space at e into two connected components. In the tangent space at e, the set of all non-vanishing vectors that are tangent to light rays consists of two connected components.*

    QL2: The set L_e of light-directions at an arbitrary event e separates the projective space at e into two connected components. In the tangent space at e, the set of all non-vanishing vectors that are tangent to light rays consists of two connected components.

Q: Since we have defined g using only points which lie in the same branch of the wavefunction as the original point e, this part of the construction is exactly the same as the classical case and thus we obtain a rank-two tensor defined on a local 'patch' all within a single branch of the wavefunction.

Just as in the previously discussed possible laissez-faire construction, we will not go through the remaining EPS axioms explicitly here, because now that we have set up the branching manifold picture together with its coordinatization, the remaining axioms do not have to be altered from the classical case.

It is worth pointing out that many of the central charges against branching spacetime formulations[27] have little force against the branching spacetime formulation of EPS *specifically*. To see this, consider the two main standard worries that concern (i) a supposed arbitrariness as to when/where the branching occurs, and (ii) violations of energy conservation. The first charge is a serious concern if one hopes to defend branching spacetimes as basic objects of general relativity (*pace* Luc, 2020), as the field equations do not determine the splitting behaviour. Indeed, not only is it unclear which branching is to be taken but also—as pointed out before—when and where.[28] However, this is not a problem in our quantum EPS construction, since the idea here is simply to provide a kinematical setup relative to which a proper dynamics is still to be formulated, and one may reasonably expect that the dynamics will determine the rules of branching, so there will not be any arbitrariness. With respect to the second charge, it is even less clear in the quantum context than in the general relativistic context as to why the satisfaction of energy conditions should be insisted upon.

[27] See Earman (2008) for a collection of objections, and Luc (2020); Luc and Placek (2020) for various counters.

[28] To be fair, Luc and Placek (2020) do provide an argument in the later part of their paper as to why branching that is not specified with respect to the when and where can be excluded on physical grounds, namely by dismissing non-Hausdorff manifolds which have bifurcating curves as non-physical. Problematically, though, these are exactly the curves to which the quantum EPS construction adheres. But, again, it is important to distinguish branching spacetime structure as basic objects of general relativity from the basic objects of some new theory, i.e. the different motivation for introducing branching spacetime structures: Luc and Placek's idea is that, for a glueing of manifolds to be sound, it should be maximal—otherwise the branching would just be arbitrary (this argument is, possibly, a good way to get around the charges of arbitrariness criticized in Luc (2020) above). Then, indeed, non-Hausdorff manifolds that involve local branching, i.e. bifurcations of the second kind, are excluded just in virtue of a theorem by Hájíček that precisely characterises such manifolds as non-maximal. But, as already explicated above, in the kinematical project of quantum EPS, it is not at all necessary to exclude branching curves to evade full arbitrariness in branching: rather, the arbitrariness in branching can be left to the dynamics, while still allowing for a local branching structure.

3.4.3 Comparison of the approaches

Both of the above laissez-faire approaches come with their own strengths and weaknesses. In the superposition-based approach (SP-EPS from now on), spacetime has already been quantized, with a Hilbert space made up of events, so we have a fully quantum spacetime at the level of the kinematics. By contrast, in the branching-based approach (BS-EPS from now on), spacetime is allowed to branch but has not been quantized, and we therefore arrive at what might be regarded as a classical branching spacetime.

One way to interpret the second approach would be to say it is based on a tacit assumption that spacetime branches are all effectively decohered gravitational histories and thus can be treated as classical objects. From this point of view, BS-EPS seems less general than SP-EPS. On the other hand, BS-EPS could still be quantized at a dynamical level—it naturally suggests an implementation of dynamics in the style of a path-integral formulation, with full flexibility as to how much interference between the different branches is to be considered.[29] Indeed, because BS-EPS does not build quantization into the kinematics, it offers greater freedom to decide later how to quantize, so in this sense BS-EPS could be regarded as being *more* general than SP-EPS. That said, because SP-EPS has already quantized spacetime, there is less work to do at the level of the dynamics, and one might hope that this would make the dynamics easier to formulate. In SP-EPS, the full branching structure (including the metric structure) formally suggests a promotion to a ket state akin to that of relativistic fields in quantum field theories.

A further interesting question concerns whether or not the two approaches are ultimately equivalent: i.e. is it the case that for any possible choice of dynamics on the SP–EPS kinematics, there is some possible choice of dynamics on the BS–EPS kinematics which will lead to an equivalent theory? The answer to this question may depend on the notion of equivalence one has in mind—it's possible that the two are *empirically* equivalent but structurally different in an important way. If the two are equivalent in the sense that one considers most relevant, then the choice between them comes down to pragmatic considerations about which one offers an easier route to a full quantum gravity theory; but if the two are not equivalent then there could be some fact of the matter about which one is a better fit to reality.[30]

[29] Another (completely different) way to make use of the kinematical structure provided by BS—EPS is to read the branches as representing Bohmian 'trajectories'. See Tumulka (2005) for a concise demarcation of Bohmian trajectories from the paths in the path-integral formulation.

[30] For an introduction to philosophical issues of theory equivalence, see Weatherall (2019a); Weatherall (2019b).

3.5 Lessons from quantum EPS

To recap: in this chapter, we have up to this point (i) reviewed the core features of the original EPS construction (§3.1), (ii) considered in some detail the main conceptual obstacle to a quantum EPS construction—namely, the notion of cross-branch spacetime point identity (§3.2), (iii) discussed various different ways in which a quantum EPS construction might be implemented (§3.3), and (iv) presented explicitly two such approaches which fall into what we have dubbed the 'laissez-faire' category (§3.4). In this final section, we turn to considering several conceptual upshots from our investigations into a quantum EPS construction.

3.5.1 Spacetime superpositions versus spatial superpositions

Consider two superficially similar cases: a light ray is put into a superposition of two different paths, and a nanoparticle is put into a superposition of two different paths. The conventional quantum description tells us that in the first case, since the difference between the gravitational fields associated with the different paths taken by the light is negligible (strictly speaking this needn't be correct, because of course electromagnetic fields still have stress-energy content, but let us simply assume that light's contribution to the gravitational field is negligible in what follows), the light goes to two different places in the same spacetime, while in the second case, because the particle is massive enough that the two different paths source significantly different gravitational fields, the massive particles end up in two different spacetimes.

However, our quantum EPS construction offers a different perspective. For suppose we take seriously the notion that points of spacetime have their identities in virtue of a construction akin to radar coordinates. Suppose that a light ray a travels in two opposite directions in two branches of the wavefunction, and that a point p is labelled in one branch of the wavefunction as the points where the light rays a, b meet. But a and b do not meet at all in the other branch of the wavefunction, so if we take identities of spacetime points to be derived from radar coordinates, it follows that the point p does not even exist in that other branch. The same goes for all other points, and so we conclude that for this kind of structuralist view about the nature of spacetime point identities, even in the case where there are no differences in spacetime structure between the branches, it nonetheless follows that different branches of the wavefunction represent different spacetimes. From this point of view, the case of spatial superpositions and spacetime superpositions are not really as different as they may first appear.

This is important for the EPS kinematics. Prima facie, one might think that we would need to distinguish somehow between the two possibilities: ordinary spatial superpositions would correspond to light rays all staying within a single spacetime, while spacetime superpositions would correspond to light rays going off into different spacetimes (or different branches of a single non-Hausdorff spacetime). But the above analysis suggests that actually we don't need to do this: any time one of our light rays splits into two separate paths, those paths are to be regarded as belonging to different spacetimes (or different branches of a single non-Hausdorff spacetime). In some cases, there will be different arrangements of matter in those spacetimes and thus we will get two distinct spacetime structures, giving us a spacetime superposition, while in other cases the arrangements of matter will be the same so we will have identical spacetime structures; but since the coupling between matter and spacetime structure comes in only at the level of the dynamics, we don't have to worry about these differences in our quantum EPS construction: kinematically speaking, the construction of spacetime is the same in each case.

Thus, on this kind of view about the nature of spacetime point identities, even when we are dealing entirely with ordinary superpositions with no different spacetime structures involved, the problem raised in §3.2 still comes into play: empirically it has been shown many times that branches of superpositions like this can indeed be recombined, but how is the map between the two branches effected? A naïve substantivalist of course can simply insist that both branches are defined on the same background spacetime, but proponents of the structuralist-type view under consideration here don't have that option, so how can they make sense of interactions between branches of a superposition at all?

One possible approach would be to deny that spacetime is constructed entirely out of the local behaviour of *actual* particles and light rays, and instead identify spacetime with the metric field as constructed out of *hypothetical* particles and light rays.[31] In this case, the direction in which the actual ray of light were to go would not matter, because the structure of the spacetime would come from the fact that hypothetically it could have gone either way: we'll get exactly the same hypothetical possibilities for the passage of light rays in both branches, so both branches of the superposition can be associated with

[31] As discussed in Chapter 1, this would be something akin to 'modal relationalism'—see Belot (2011).

the same background spacetime, which tells us how to map between the two branches. However, one might question whether this modal take on spacetime and metric structure is strictly in accordance with the operationalist and/or relationalist underpinnings of EPS.

Note also that if we say that the existence of identical (modal) spacetime structure is essential to the possibility of mapping between the branches in the spatial superposition case, then it would seem to follow that there can be no such map in the spacetime superposition case, where we no longer have identical spacetime structure. Thus, if we want to maintain the existence of spacetime superpositions (and the possibility of recombination for such superpositions), we can't insist that the identical structure plays any special role here; and in any case since spatial superpositions can be regarded as simply examples of spacetime superpositions in the limit as some measure of difference of structure goes to zero, we might naturally hope that the procedure for mapping points between branches should be roughly the same in each case—so we should use something like best-matching even in the spatial superposition case (we suggest that Barbour himself would probably advocate this).

3.5.2 Comparison with other cases for the quantum nature of gravity

In the process of constructing a quantum version of EPS, we saw that by starting with some operationally legitimate assumption about quantum signals, we arrive at the conclusion that spacetime structure involves a superposition of spacetimes in one form or the other. In other words, we have thus arrived at yet another plausibility argument as to why gravity should be quantum. In this section we would like to consider how far the plausibility argument for the quantum nature of gravity differs from by-now familiar ones, which are, *inter alia*, arguments from (i) analogy (say, to electrodynamics), (ii) inconsistency of semi-classical gravity, and (iii) inconsistency of classical gravity with gravitational-induced entanglement.

More specifically, the quantum EPS argument (QEA) argument goes as follows. Let us understand the EPS construction as a realization of the relationalist vision of constructing spacetime out of the behaviour of light and matter; then we argue that since light and matter do different things inside different branches of the wavefunction, different spacetimes necessarily get constructed inside these different branches, so we get superpositions of spacetimes and therefore gravity must be quantized. That is, the quantization of gravity follows directly from the supposition that it is legitimate to construct spacetime out of the behaviour of light and particles, as the original EPS construction does.

Of course, like all arguments to the effect that gravity must be quantum, this argument is based on some assumptions: primarily the relationalist assumption that spacetime is in some sense a codification of the behaviour of light and matter, or at the very least (if for whatever reason one is not willing to accept relationalism) the idea that there is necessarily a tight link between what to count as the structure of spacetime and the behaviour of light and particles. Evidently the assumptions of QEA are more philosophical than the assumptions used in other plausibility arguments, which mostly employ specific physical conjectures—for example, the argument of Eppley and Hannah (1977) (an argument of type (iii)) requires the assumption that non-gravitational measurements lead to a wavefunction collapse. This comparison makes it clear that QEA depends on pre-existing philosophical prejudices in ways that the other arguments do not, and therefore QEA will not seem compelling to someone who is not inclined towards the relationalist position in the first place. On the other hand, because QEA argues for the quantization of spacetime purely on the basis of a conviction about the nature of spacetime, it is less vulnerable to refutation by the proposal of alternative models which don't have the physical features which led to inconsistencies in the original model (as, e.g. has happened in the case of the Eppley–Hannah argument (Huggett and Callender, 2001; Mattingly, 2006)). So for those who do favour the relationalist view, QEA may seem a more robust argument than some of the alternatives.

QEA also stands out in virtue of the fact that it is kinematical and constructivist, whereas most of the other arguments for the quantum nature of gravity take place at the level of the dynamics. That is to say, since our quantum EPS construction has focused entirely on constructing the space of possible states rather than defining any specific dynamics on that state space, the resulting argument for the existence of spacetime superpositions is not dependent on details of the dynamics. Of course, it is true that the nature and extent of the spacetime superpositions that we would expect to see in an actual case depends on dynamical features; most importantly, it's likely that in order to get distinguishably different spacetimes in different branches one would need to have quite large numbers of particles in distinct states in those branches, so the actual impact of spacetime superpositions on physically observable processes will be shaped by dynamical factors like decoherence. But our constructions do stipulate that in principle we get branching spacetimes even if the number of particles involved is small, so while the construction does not itself guarantee that spacetime superpositions will be widespread, significant or even observable, it does immediately have the consequence that such things can exist. This, again, makes QEA more robust than some alternatives, as it does not depend on any details of the dynamics and thus will be valid for a large variety of choices for the dynamics. However, in virtue of being kinematical the QEA does run into all of the ambiguity issues that we have discussed in this chapter. But if the kinematicist line of the QEA is given up, it is hard to see how one can still be constructivist, which makes up much of the bite of the QEA. To understand the situation better, it is helpful to recall that the classical EPS scheme can, in virtue of its (more or less consequent) (empirical) constructivist nature, show compellingly at what point the phenomena allow for a much more general kinematical arena than Lorentzian geometry—think for instance of how neither torsion nor differentiability of the metric are backed up well from data or other principles, leaving theories such as teleparallel gravity and Finsler spacetime as live options. Similarly then, the quantum EPS scheme, by being constructivist, despite—or perhaps *because of*—its difficulties, makes us aware of the intricacies of formulating quantum general relativity and neighbouring theories. So, as long as we do not think of the constructivist undertaking in any strict epistemic or metaphysical fashion but more as an auxiliary tool for theory construction, we can accept issues of circularity and happily appreciate instances of ambiguities (such as may arise with the kinematicist commitment).

Finally, it is worth comparing the scope of the arguments: QEA establishes the quantum nature of gravity in relation to relatively high-level quantum particles or 'light rays' that themselves do not have internal structure: the treatment of light just happens at the level of ray optics. In a sense, maybe this limitation is a virtue, as the QEA can thus be seen as an argument for quantum gravity even at rather low energies, such as those studied in low-energy perturbative quantum gravity. Note in this context that arguments of type (i) to (iii) also only establish the quantum nature of gravity for a limited scope; in particular, they say little about what quantum gravity (if one still even would like to call it such) would look like at lower energies.[32]

[32] For instance, with respect to (i), the analogies to electrodynamics, as well as the identification of gravity by analogy as yet another spin-n, prima facie only establish a quantum theory of gravity in line with (non-fundamental) quantum electrodynamics and (quantum) effective field theory, respectively.

3.5.3 Quantum gravity

It will also be useful to consider how the quantum EPS construction we have discussed in this chapter might be related to various actual attempts to arrive at a quantum theory of gravity. First of all, we do not think it is likely that a constructivist approach in and of itself should be regarded as a theory of quantum gravity. The key problem here is that anything which can reasonably be called 'contructivist' must be based on axioms which are at least somewhat operational in character—of course this line is somewhat blurry, but for example, as noted in §3.3.2, it seems reasonable to say that axioms involving causal diamonds are acceptable whereas axioms involving edges of a LQG graph are not. Yet the domains in which quantum gravity effects become most significant are typically very far removed from any remotely operational interpretation—e.g. we expect such effects to be important at the Planck scale, in the very early universe, or at a black hole singularity, all regimes not expected to be well-populated with macroscopic observers. Moreover the absence of observers in these regimes is not merely an unlucky contingency; it seems very likely that there *could not possibly be* any macroscopic observers in these regimes, and thus we can't even make counterfactual claims about the operationally defined phenomena that observers *would* see if they *were* present in such regimes. Thus it would arguably be quite surprising if some constructivist approach were able to completely pin down the physics of such regimes; in particular, our usual way of thinking about relationships between scales suggests that the operational phenomena observable at macroscopic scales should supervene on the quantum-gravitational phenomena defined at the Planck scale and not vice versa, which would mean in effect that there is a one-to-many map from operational phenomena to Planck-scale effects, so the operational phenomena cannot completely fix the Planck-scale effects (at least, not without subtle experiments probing highly specific microscopic phenomena, whose results would not necessarily be predictable within any standard constructivist picture based on generalizing ordinary spatiotemporal experiences).

Nonetheless, we think this exercise of seeking an EPS-style constructivist account of quantum gravity may still usefully inform ongoing work on the 'real' problem of quantum gravity. One key insight here is that even if we are only seeking to establish a high-level, essentially 'macroscopic' kinematics for quantum gravity, there are still a number of choices to be made and challenges to be overcome; in particular, all of these different options appear to be compatible with all existing empirical evidence and our common-sense understanding of spacetime, so there is no obvious way to single out a specific correct approach. Of course, one might think the right thing to do here is simply to put aside the macroscopic kinematics and work directly on the putative underlying Planck-scale physics, with the expectation that the Planck-scale physics may eventually answer our questions about the macroscopic kinematics; but the danger of that strategy is that we may be implicitly making some assumptions about the appropriate macroscopic kinematics and allowing that to shape the direction of our research on the Planck-scale physics, without adequately considering that we could have taken a different path right from the beginning by choosing a different approach to macroscopic kinematics. In addition, we emphasize that right now, some approaches to quantum gravity do not deliver unambiguous answers to the kinds of questions about macroscopic kinematics that we have raised here. For example in LQG it is not yet clear how to define the classical macroscopic limit in the most general case, and indeed the easiest way to extract predictions from the theory is to use it to calculate transition amplitudes for a classical boundary, with quantum-gravitational effects confined to the interior of the boundary (Rovelli and Vidotto, 2022). This scenario doesn't require an answer to the questions we have posed here, since 'operational' phenomena are confined to the boundary region defined on a spacetime which is typically either fully classical or at least fully decohered; meanwhile the quantum gravity effects inside the region are used only to predict the behaviour at the boundary, so no concrete story is offered about what is happening to spacetime inside the region, and therefore don't need to specify what a superposition of spacetime looks like or how it could get recombined. But ultimately if LQG is to be successful, one would expect that it should also be able to reproduce situations like the BMV experiment involving macroscopic or mid-sized superpositions, and thus in the end it must offer some kind of answer to the questions we have been concerned with here. Thus the constructivist approach is potentially useful both for setting out precisely the kinds of questions we would like a theory of quantum gravity to answer and also for more clearly delineating the range of options.

3.5.4 The quantum hole argument

We close by considering how the issues we have discussed in this chapter may give new insight into the hole argument. To motivate this, we recall Penrose's argument against spacetime superpositions which was introduced in §3.2.2, and we suppose for the moment that Penrose is correct: that is, *if* there exists no privileged mapping of one superposed spacetime to the other, then no time evolution operator can be defined and therefore the superposition must undergo a gravitationally induced collapse. On the other hand, if spacetime points have primitive identities, then there does exist a privileged mapping of one superposed spacetime to the other, and then in turn a time evolution operator *can* be defined and therefore presumably we will not observe any gravitationally induced collapse. If this argument were correct, then this would be an instance where the existence or nonexistence of primitive identities for spacetime points would have an immediate and in principle observable impact on real physics: the existence of primitive identities gives us a stable state, whereas the absence of primitive identities leads to instability and collapse.

Now we don't in fact think Penrose's argument is correct, because we agree with Giacomini and Brukner (2021) that there is no reason to think that a spacetime superposition must contain a unique time evolution operator which applies across all the branches of the superposition. However, Penrose's argument nonetheless illustrates the fact that questions around the identity of spacetime points take on a new relevance when we begin to consider the possibility of quantum diffeomorphisms (as introduced in §3.2). The usual form of the hole argument requires us to make comparisons across different possible worlds, and thus whilst these sorts of arguments may have some indirect empirical consequences (e.g. they may motivate us to impose a diffeomorphism constraint) there is of course no possibility of directly observing relationships between different possible worlds. On the other hand, we can certainly devise observations which 'observe' happenings in different branches of the wavefunction, or tell us something about the relationships between different branches of the wavefunction—e.g. this is the goal of various proposals to detect quantum gravitational effects using tabletop experiments, such as Feynman's early ideas about detecting superpositions of spacetimes by means of interference between the different branches, or the more sophisticated BMV experiment discussed earlier in this chapter (Bose et al., 2017; Marletto and Vedral, 2017). These experiments would not be easy to perform but seem at least in principle possible, so it is at least conceivable that if spacetime superpositions exist, then questions about the identities of points across branches may have direct significance for real observations. So the upshot of our discussion here is that introducing quantum phenomena is likely to make hole argument-style concerns more pressing and empirically relevant: the indeterminism at the centre of the argument becomes not merely a matter of which points instantiate which field values, but rather a matter of there being empirically discernible differences.[33]

But the approach we have adopted here also suggests a way of avoiding the hole problem. For in the EPS context, one might hope that the use of radar coordinates should prevent the hole argument from being made. For radar coordinates label spacetime points in terms of their relation to particles and light rays, and since particles and light rays are ultimately made out of fields, presumably when we perform a diffeomorphism on all the fields as in the hole argument, those particles and light rays will be moved along with the diffeomorphism, meaning that the coordinates will also be moved by the diffeomorphism, and therefore the relation between the fields and the coordinates will not change during the diffeomorphism. Thus, it might seem that the radar coordinate approach is a particularly appropriate way of assigning coordinates in the context of general covariance—not only the physics but even the coordinatization will be invariant under diffeomorphisms. However, this strategy will work only if the radar coordinates are defined with respect to the behaviour of *actual* particles and light rays. If on the other hand we regard the particles and light rays as purely mathematical objects which are used to derive some abstract choice of atlas for the point set, then since the diffeomorphism used to define the hole argument acts only on actual fields and not hypothetical ones, these radar coordinates will not be moved by this diffeomorphism, and thus we will again encounter the hole problem. One could of course adopt more complex identity conditions for spacetime points so as to impose the requirement that hypothetical as well as actual fields are shifted by diffeomorphisms, but that would amount to a different kind of solution to the hole problem, in the spirit of sophisticated substantivalism (Pooley, 2005; Hoefer, 1996). So if we want to use the radar coordinate scheme as a novel way of avoiding the hole problem, it is important that the light rays and particles should be actual.

There are a number of obstacles to this way of avoiding the hole problem. First, in order for actual light rays and particles to give rise to a complete, manifold-like coordinatization of the point set, it seems that both particles and light rays would have to be fairly ubiquitous, in order that every point should be reached by enough light rays to ensure that it gets radar coordinates. It's possible that our actual universe does indeed have a sufficiently large number of particles and light rays to do this (e.g. if we make use of the cosmic

[33] Cf. Pooley and Read (2021).

microwave background), but on the other hand this doesn't seem guaranteed. And making this requirement imposes a constraint for the electromagnetic matter sector from which light rays derive: as light rays result from the high-frequency limit of propagating electromagnetic waves, electromagnetic field tensors (and, consequently, stress–energy–momentum tensors) will only be acceptable if they lead to sufficient light rays upon that limit. In particular, a strictly global electrostatic (as opposed to electrodynamic) field content in the universe would not be sufficient to allow for the existence of actual radar coordinates. Thus if this route is adopted, EPS must be expanded to a programme for establishing the right kinematics for the Einstein–Maxwell theory rather than just GR.

Second, we note that there is a clash between taking the light rays in the radar coordinates to be actual, and the proof strategy of classical EPS: for the proof relies heavily on the notion that the radar coordinates establish a chart on a manifold and that, once they have done so, charts other than radar coordinates can be used. One could perhaps argue that other coordinate systems can still be used as mathematical coordinates even if it is the radar coordinates which are taken seriously as a meaningful physical labelling of the points, but it would need to be shown explicitly that the proof strategy of EPS still goes through under this interpretation.

Finally, there is not just one possible radar coordinate system: the radar coordinates depend on the two particle trajectories with respect to which they are defined, and each set of radar coordinates is of limited scope. But given that a compatibility condition needs to hold between radar coordinate systems, one may suspect that the radar coordinate change should be regarded as a passive transformation as between any coordinate system change on a manifold, thus inducing also a corresponding active transformation which will necessarily revive a hole-argument scenario (as already alluded to above). However, to just apply the notion of passive and of active transformation as familiar from the manifold context (and criticize the physical reading of radar coordinates in terms of actual light ray signals) amounts to a *petitio principii*. The real question here is whether one can—and, perhaps also, whether one even needs to—formally express the radar coordinate charts in a way that prevents equivocation with standard manifold charts. One proposal could consist of piggy-backing on the manifold structure: we could simply acknowledge the reciprocal dependence of the charts on the the fields they are describing. However, what about the one-dimensional line of the observer? The radar coordinates are characterized in terms of the emission and the arrival times, which are—on the standard EPS approach—values in one-dimensional coordinate charts for the two defining particle of the radar coordinates.[34] Can we even think of these times as dependent on the field content?[35]

3.6 Outlook

This investigation into the possibility of a quantum generalization of the EPS construction has given rise to many interesting questions. First, since the EPS approach constructs only kinematics, one might naturally ask whether the *dynamics* of a quantum theory of gravity can be constructed on top of the already established kinematical structure. Recall for a moment the situation in classical EPS, in which the dynamics—the Einstein field equations in the case of GR—is obtainable from a best-system analysis, i.e. read out as the best codification of how the kinematical structure evolves in different contexts (say by tracking what kind of energy–momentum content—or even, more directly, what kind of matter field values—corresponds to what kind of values of the basic kinematical variables established by EPS). If such a proposal already sounded unrealistic in the classical case, how is it ever to be practically implemented in the quantum case, where we can only observe one out of multiple branch structures? After all, in the quantum case, data points have to be collected across different contexts as well as histories within one and the same overall context ('branches'). But note that, at the end of the day, the project of setting up a quantum EPS construction is not one of literally arriving at a proposal for a theory of quantum gravity but one of conceptual clarification—with the major lessons summarized in the previous section.

In any case, it seems necessary to say more on how exactly the dynamics is expected to dispel the aforementioned threat of arbitrariness with respect to when and where exactly curves branch—and when and where they do not. One simple option is to render gravitational branching as induced by matter branching: we are familiar with how matter branching comes about (e.g. why a beam splits in the interaction with a beam splitter); if the splitting of a manifold is simply—and only—due to matter splitting, and if—as standardly seen—the matter splitting is not arbitrary, then there is no longer any issue with arbitrariness in curve branching either. Admittedly, though, this resolution

[34] Axiom D_1 has it that 'Every particle is a smooth, one-dimensional manifold'.

[35] In fact, how to think of the emission and arrival time is a general issue worth further consideration, even if one takes radar coordinates to be on a par with regular coordinates: what is required, and what exactly is supposed to give rise to the observer's emission and arrival time? In what sense is EPS not after all committed to local clocks?

only works for a lower-energy regime of general relativity for which gravitational effects are fully determined by the matter theory (see Anastopoulos et al. (2021)). For the actual regime of quantum gravity which is decisively marked by independent gravitational degrees of freedom, the issue will be more cumbersome.

3.A General covariance and quantum diffeomorphisms

As mentioned in the main text of this chapter, Anandan (1997), Hardy (2019), Giacomini and Brukner (2021), and others have all proposed some variant or other of a principle of 'quantum general covariance', which encodes invariance under quantum diffeomorphisms. Note that these authors seem to be equating diffeomorphism invariance with general covariance, though in fact, following Pooley (2017), there is a case to be made that the two notions should not be regarded as being identical. Some unpacking is thus in order.

Various different definitions have appeared in the literature, but following Pooley (2017, pp. 115–7), let us say that general covariance is the requirement that 'the equations expressing [the] laws are written in a form that holds with respect to all members of a set of coordinate systems that are related by smooth but otherwise arbitrary transformations', while diffeomorphism invariance is the requirement that 'if $\langle M, F, D\rangle$ is a solution of the theory, then so is $\langle M, F, d_* D\rangle$ for all diffeomorphisms d' (here, M is the background manifold, F are any solution-independent fixed fields on the M, D are dynamical fields, and stars indicate push forwards[36]). Famously, as pointed out by Kretschmann, the requirement of general covariance is not really a restriction on possible theories at all: even pre-relativistic theories and special relativity can be put in a generally covariant form by simply turning the background metric into a mathematical object which features explicitly in the laws of the theory.[37]

In contrast, diffeomorphism invariance is a stronger requirement: for example, a generally covariant formulation of special relativity where the Minkowski metric is made explicit in the form of a solution-independent fixed field will fail to be diffeomorphism-invariant, since applying a diffeomorphism to the dynamical fields but not the metric will move the fields around relative to

[36] For more on what we take Pooley to mean by a 'fixed field', see Read (2020b).

[37] For more on the history of general covariance and Kretschmann's objection, see Norton (1993). The debate over general covariance has since transformed into a debate over the 'background independence' of general relativity: for further discussion of this issue, see Pooley (2017) and Read (2023b).

the metric and thus will typically give rise to a model which is not dynamically possible. However, in most cases it is still possible to find a way of making a theory invariant under diffeomorphisms: we simply have to turn the fixed fields into dynamical fields. For example, in the case of special relativity we could do this by making the metric a dynamical object governed by an equation requiring that its Riemann curvature tensor is zero everywhere, so now diffeomorphisms will act jointly on the metric and the other fields and will thus take solutions into solutions.

Similar points can be made in the quantum case. By analogy with the classical case, let us say that quantum general covariance is the requirement that

> the equations expressing the laws are written in a form that holds with respect to all members of a set of coordinate systems that are related by arbitrary quantum diffeomorphisms.

As in the classical case, this requirement is not very substantive: even if the predictions of a theory depend non-trivially on an identity map between branches, the theory can be still put in a generally covariant form if we turn the identity map between branches into a mathematical object which features explicitly in the laws of the theory. Similarly, let us say that invariance under quantum diffeomorphisms is the requirement that

> if $\langle M, F, D\rangle$ is a solution of the theory, then so too is $\langle M, F, d_* D\rangle$, for all quantum diffeomorphisms d.

This is a stronger requirement than quantum general covariance, but again, even models with non-trivial dependence on an identity map between branches could be put in such a form, provided we can find a way of making the identity map at least nominally dynamical by writing it as the solution to some equation.

One way to make the requirement of diffeomorphism invariance more substantive in either the classical or the quantum case is to be more strict about what counts as a dynamical field. For example, inspired by Einstein's action–reaction principle, one might argue that any genuinely dynamical object should not only act but also be acted back upon.[38] Under that stipulation, it would follow that in a special relativistic theory the metric would not really

[38] See Brown and Lehmkuhl (2016) for recent discussion of Einstein's understanding of the action-reaction principle.

be dynamical, meaning that it must be regarded as a fixed field, and therefore special relativity would fail to exhibit diffeomorphism invariance. Similarly, in the quantum case, if the identity map between branches is independent of the behaviour of the matter in the branches, then it would have to be regarded as a fixed field, so any model where the predictions depend non-trivially on the identity map would fail to exhibit diffeomorphism invariance. Under this construal, in order to have invariance under quantum diffeomorphisms, we would either have to ensure that the identity map is purely a form of gauge, or we would have to make the identity map depend on the configuration of matter—e.g. by defining an identity map in terms of similarity of matter content. Another way of thinking about general covariance is due to Barbour (2001), who argues that the true empirical content of general covariance is the manner in which the respective four-dimensional objects are assembled out of three-dimensional constituents. If we accept Barbour's argument, then in fact 'quantum general covariance' should entail not that the theory is invariant under quantum diffeomorphisms, but rather that it uses identities between spacetime points in different branches which are determined by best matching and not by some other absolute structure.[39]

[39] For further discussion of this proposal and others, see Read (2023b).

4

Non-relativistic Constructive Axiomatics

Introduction

In the previous chapter, we saw how one might modify the axioms of the 1972 constructive axiomatization of general relativity due to Ehlers, Pirani, and Schild (EPS) so as to better model quantum mechanical empirical phenomena. In this chapter, we continue in the same spirit, by considering how the EPS axioms might be modified so as to be suited to non-relativistic inputs. Further, we show that by modifying the notion of relativistic conformal structure (in infinitesimal form: a field of cones over the manifold) to be appropriate for the non-relativistic context (effectively by 'widening' the cones as one increases the speed of light), and by building up said structure from elementary and empirically informed axioms *à la* EPS, one can arrive at the structure of a non-relativistic spacetime (a 'classical spacetime', to use the terminology of (Malament, 2012, ch. 4)) and (thereby) set this spacetime on more secure empirical footing.[1]

Of course, of any such project, one might reasonably ask: *why bother*, especially when we (think we) know that the universe is in fact relativistic in nature? To this question we offer three responses. First, there are certainly regimes in which the velocity of light is *effectively* infinite; one might wonder which spacetime structure is yielded by the causal–inertial constructive axiomatic approach in this case, since said spacetime will reasonably qualify as the *effective* spacetime structure in this regime.[2] And related to this: one must remember that 'true'—i.e. strong field—general relativity is rather far removed from the empirical anyway. Therefore, the kind of observations we would expect to make as local observers should be at most corrections to the Newtonian picture. Second, the work of the foregoing chapters has

[1] This approach to a non-relativistic version of EPS is quite natural, if one takes a 'geometric' understanding of non-relativistic spacetime physics on which said physics is obtained by widening light cones—see Fletcher (2019) for a detailed discussion of the approach. If one has an alternative understanding of non-relativistic gravity according to which there simply is no 'law of light' prohibiting superluminal propagation, then whether light (or other signals) have finite or infinite propagation speed will depend upon the details of the dynamics involved.

[2] Cf. (Wallace, 2020).

Constructive Axiomatics for Spacetime Physics. Emily Adlam, Niels Linnemann, and James Read, Oxford University Press. © Emily Adlam, Niels Linnemann, and James Read (2025). DOI: 10.1093/9780198922391.003.0005

already (we hope!) demonstrated that one secures a better handle on the architecture of a general relativistic spacetime—and the possibilities for modifying said spacetime—when one understands how that spacetime is built up from sub-metrical constituents such as projective and conformal structures. In a similar manner, one motivation for undertaking the work of this chapter is that, having done so, we will thereby come to understand better how the structures of a classical (i.e. non-relativistic) spacetime are built up, and (again) how those structures might, in principle, be modified. And third: in light of the recent development of 'type II Newton–Cartan theory', it is not so obvious that the universe *must* be relativistic in nature, given that said theory can (it seems) recover many of the empirical predictions of GR.[3]

Setting aside these worries, then, our general agenda for this chapter has three items:

1. Provide definitions of non-relativistic projective/conformal structures in both 'infinitesimal' and 'algebraic' forms (cf. §1.1), and demonstrate the equivalence of those forms.
2. Demonstrate that non-relativistic projective and conformal structures can be built up from elementary, empirically informed axioms *à la* EPS.
3. Build up from non-relativistic projective and conformal structures the structure of a classical spacetime, following the methodology of Matveev and Scholz (2020) as presented in the relativistic case.

As such, our plan for the chapter is this. Each of (1)–(3) above will be discussed in, respectively, §4.2–§4.4. Before doing so, however, we review in §4.1 some previous work which has been undertaken on these issues. We close in §4.5 with some philosophical discussion.

4.1 Previous works

Perhaps not unsurprisingly, there is very little existing literature addressing the question of whether the EPS approach can be modified so as to be appropriate for the non-relativistic setting. One important exception is Ewen and Schmidt (1989). On reading that article, however, one might quite reasonably

[3] For general background on type II Newton–Cartan theory, see Hansen et al. (2020); for discussion of this theory in the context of tests of general relativity, see Hansen et al. (2019); Wolf et al. (2023b).

be confused as to what exactly its authors achieve. In the abstract, its main result is summarized as follows:

> It is shown that the differentiable, affine, and metric structure of Newton-Cartan space-time is uniquely determined by its projective-conformal-material structure. (Ewen and Schmidt, 1989, p. 1480)

This, therefore, makes it sound as though the authors have proved a non-relativistic version of Weyl's uniqueness theorem: a Newton–Cartan spacetime (by which the authors here mean a classical spacetime—nothing of the specific dynamics of Newton-Cartan theory need yet be invoked) is fixed uniquely by its 'projective-conformal-material' structure (we will return later to what 'material' means here). On the other hand, later in the same abstract its achievement reads somewhat differently:

> This means physically that—similarly as in general relativity—it is also possible in classical gravitation to define operations of parallel transport and measurements of length, time, and mass using only three kinds of world lines: world lines of freely falling test particles, of photons, and of gravitational matter. (Ewen and Schmidt, 1989, p. 1480)

Evidently, there is some equivocation here, because this second passage makes the achievement of the article sound like an *existence* proof *à la* EPS: given a 'projective-conformal-material' structure, a Newton-Cartan spacetime structure is thereby fixed.

So, which one is it? Do Ewen and Schmidt (1989) prove the uniqueness result or the existence result (if either)? Confusion here is deepened by the fact that, in the introduction to their article, the authors attribute incorrectly to EPS the Weyl-style uniqueness result; for this reason, they take themselves to be offering a non-relativistic version of EPS. The correct way in which to read their result, however, is this: they do indeed prove a non-relativistic version of Weyl's uniqueness result, but incorrectly assimilate Weyl's result to EPS' achievement. As such, the project of providing a non-relativistic version of the EPS existence result, and (moreover) of building up non-relativistic projective and conformal structures from elementary axioms, remains outstanding; it is our goal in this chapter to (*inter alia*) achieve these tasks for (what is to our knowledge) the first time.

To close this section: it is incumbent upon us to mention two other pieces of work which are relevant to our project in this chapter. Quite independently of

Ewen and Schmidt (1989), Curiel (2015) also proves a non-relativistic version of Weyl's uniqueness theorem—i.e. proves that a classical spacetime structure is fixed uniquely by its associated non-relativistic projective and conformal structures.[4] Interestingly (and importantly), Curiel's notion of non-relativistic conformal structure is in fact not equivalent to that used by Ewen and Schmidt (1989); the differences are reviewed by March (2023), and we discuss them ourselves below.

4.2 Non-relativistic projective and conformal structures

Having now reviewed the existing literature on this topic and seen clearly that there remains a gap in the literature for a non-relativistic version of EPS, we turn in this section to providing both 'algebraic' and 'infinitesimal' versions of non-relativistic projective and conformal structures, and to proving—in the spirit of §1.1—the equivalence of these formulations. Then (to repeat), we will in §4.3 demonstrate that, in the spirit of EPS, these pieces of non-relativistic spacetime structure can be built up from elementary, empirically informed axioms. Finally, in §4.4 we will prove that those two structures (plus a 'material' structure) fix a classical spacetime structure (thereby proving a non-relativistic version of the EPS existence result).

4.2.1 Projective structure

Our starting point is to identify the correct non-relativistic analogues of projective and conformal structures. We'll begin with projective structure. Recall from §1.1 that this structure can be presented in both 'algebraic' and 'infinitesimal' versions as follows:

Algebraic: An equivalence class of affine connections $\Gamma, \Gamma', \ldots$ on the manifold M given as follows:

$$\Gamma \overset{\mathcal{P}}{\equiv} \Gamma' \quad \Leftrightarrow \quad \exists \text{ a 1-form } \psi \text{ on } M \, s.t. \, \Gamma'^{\mu}_{\nu\rho} = \Gamma^{\mu}_{\nu\rho} + \delta^{\mu}_{\nu}\psi_{\rho} + \delta^{\mu}_{\rho}\psi_{\nu}.$$

Infinitesimal: A family of curves, called *geodesics*, whose members behave in infinitesimal neighbourhood of each point $p \in M$ as straight lines in projective four-space.

[4] As Curiel himself acknowledges (p.c.), there is an error in the pre-print version of Curiel (2015): one in fact requires a 'material' structure *à la* Ewen and Schmidt (1989) in order to complete the proof. We will return to this point later in the chapter. (Interestingly, Curiel also seems to incorrectly assimilate the EPS result to Weyl's uniqueness theorem—see Curiel (2015, fn. 2).)

Note now, though, that nothing here is specific to a relativistic spacetime setting! (This, indeed, is recognized by both Ewen and Schmidt (1989) and Curiel (2015).) Therefore, we can in fact continue to work with this notion of projective structure as we explore non-relativistic versions of EPS; no modifications are required.

4.2.2 Conformal structure

Finding a suitable non-relativistic analogue of conformal structure is a substantially more delicate business. Let us again begin by recalling the previous definitions—both 'algebraic' and 'infinitesimal'—of conformal structure in the relativistic setting:

Algebraic: An equivalence class of metric fields $g, g', \dots$ on M given as follows:

$$g \overset{\mathcal{C}}{\equiv} g' \quad \Leftrightarrow \quad \exists \text{ a function } f \text{ on } M \text{ s.t. } g' = e^f g.$$

Infinitesimal: A field of infinitesimal null cones defined over M.

These definitions clearly will not do in non-relativistic contexts, for (thinking in terms of the algebraic characterization) one does not have Lorentzian metrics $g_{\mu\nu}$ on the manifold in the relativistic case (rather, one has pairs of degenerate temporal metrics t_μ and spatial metrics $h^{\mu\nu}$ which are orthogonal to each other, along with a derivative operator ∇ which is compatible with both t_μ and $h^{\mu\nu}$—see Malament (2012, ch. 4)). Moreover (thinking in terms of the infinitesimal characterization), one does not have null cones in the non-relativistic setting, for in this case there is no upper limit to the velocity of propagating signals.

We'll return in a moment to the infinitesimal characterization; for now, let us focus upon the algebraic characterization. Here, there is no fixed consensus as to the most appropriate definition of this notion. Let's begin with the definitions provided by, respectively, Ewen and Schmidt (1989) and Curiel (2015); here, for simplicity, we make use of the elegant and compact statements of these notions given by March (2023):

Conformal-ES: Let $\mathcal{M} = \langle M, t_\mu, h^{\mu\nu}, \nabla\rangle$ be a non-relativistic spacetime. The Ewen–Schmidt conformal structure $\mathcal{C}_{\mathrm{ES}}$ of $\mathcal{M}$ is the set $\{\sigma\}$ of unparameterized spacelike geodesics of ∇.

Conformal-C: Let M be a differentiable four-manifold. A Curiel conformal structure $\mathcal{C}_{\mathrm{C}}$ on M is an integrable rank-3 distribution on M with leaves S, and an equivalence class of conformally equivalent spatial metrics $[h^{\mu\nu}]$ on M such that the following both hold: (a) there exists a one-form t_μ on M, orthogonal to all the $h^{\mu\nu} \in [h^{\mu\nu}]$ such that for all points p, all S, and all tangent vectors σ^μ to S at p, $t_\mu \sigma^\mu = 0$, and (b) at least one representative of $[h^{\mu\nu}]$ is flat.

A couple of points here are already in order. First, the definition of $\mathcal{C}_C$ is given only for four-manifolds—however, the extension to manifolds of arbitrary dimension should be self-evident (one simply considers a codimension-1 distribution on $\mathcal{M}$) and as such we will not dwell on it further. Second, as March (2023) points out, Curiel (2015) in fact defines separately notions of spatial and temporal conformal structure for a non-relativistic spacetime and then demands compatibility of the two (for a certain sense of 'compatibility'). This is innocuous in the sense that in the spatially flat spacetimes which Curiel considers the two are equivalent. Third: following March (2023), one might wish to generalize $\mathcal{C}_C$ by removing condition (b) in its definition. This leads to the following notion of a non-relativistic conformal structure—it is this structure which we will refer to in what follows as 'non-relativistic conformal structure':

Conformal-NR: Let M be a differentiable four-manifold. A non-relativistic conformal structure $\mathcal{C}_{\mathrm{NR}}$ on M is an integrable rank-3 distribution on M with leaves S, and an equivalence class of conformally equivalent spatial metrics $[h^{\mu\nu}]$ on M such that there exists a one-form t_μ on M, orthogonal to all the $h^{\mu\nu} \in [h^{\mu\nu}]$ such that for all points p, all S, and all tangent vectors σ^μ to S at p, $t_\mu \sigma^\mu = 0$.

Let's return for a moment to the difference between $\mathcal{C}_{\mathrm{ES}}$ and $\mathcal{C}_{\mathrm{C}}$. March (2023) summarizes these as follows:

> Given the Ewen-Schmidt conformal structure [...] we can define a foliation of M into spacelike hypersurfaces (since, up to conformal factors, there will be a unique one-form which annihilates all tangent vectors to all the $\sigma \in \mathcal{C}_{\mathrm{ES}}$). But $\mathcal{C}_{\mathrm{ES}}$ gives us more than this: we can also recover a notion of spatial projective structure—i.e. the class of spatial derivative operators D induced on each spacelike hypersurface for which the $\sigma \in \mathcal{C}_{\mathrm{ES}}$ can all be reparameterised as geodesics.
>
> By way of contrast, the Curiel conformal structure [. . .] also gives us a foliation of M into spacelike hypersurfaces; unlike $\mathcal{C}_{\mathrm{ES}}$, however, $\mathcal{C}_{\mathrm{C}}$ does not fix the projective structure of the leaves of the foliation, instead, it determines a conformal structure on each leaf. So whilst $\mathcal{C}_{\mathrm{ES}}$ and $\mathcal{C}_{\mathrm{C}}$ are not orthogonal, neither is strictly weaker than the other. Moreover, the fact that $\mathcal{C}_{\mathrm{ES}}$ partially determines the projective geometry of M suggests that it is $\mathcal{C}_{\mathrm{C}}$, rather than $\mathcal{C}_{\mathrm{ES}}$, which is the closer analogue of relativistic conformal structure for the non-relativistic context. (March, 2023, p. 6)

We concur completely with this assessment; as such, it is with $\mathcal{C}_C$—or, rather, with its generalization $\mathcal{C}_{NR}$—which we will work going forward. Importantly, however, note that at this point we have presented only an *algebraic* characterization of non-relativistic conformal structure, for (as with the Ewen–Schmidt and Curiel characterizations), it makes appeal to equivalence classes of as-yet unconstructed pieces of spacetime structure, which runs contrary to the constructivist pretensions of EPS. Thus, what we seek now is an equivalent, 'infinitesimal' characterization of $\mathcal{C}_M$. What would this look like?

The answer here is not difficult to provide. Here, to be specific, is our proposal:

Conformal-NR (infinitesimal): Let M be a differentiable four-manifold. A non-relativistic conformal structure (characterized infinitesimally) $\mathcal{C}_{NR}^{\text{inf}}$ on M is an integrable rank-3 distribution on M with leaves S, giving a stipulation of facts about angles between any two intersecting worldlines on any given leaf $S \subset M$, such that there exists a one-form ζ_μ on M such that for all points p, all S, and all tangent vectors σ^μ to S at p, $\zeta_\mu \sigma^\mu = 0$.

There are a few points to note here. The fact that the manifold M is foliable into hypersurfaces is a topological restriction which itself guarantees that there exists some 'topological time' ζ parameterizing the 'distance' between leaves of the foliation (Chen and Read, 2023, p. 5).[5] From ζ, one can then define $\zeta_\mu := d_\mu \zeta$, where d denotes the exterior derivative on M. In other words, some ζ_μ is definable from the integrable rank-3 distribution on M with leaves S. Second: the notion of a tangent vector presupposes no metrical structure, and therefore is acceptable from the point of view of characterizing 'infinitesimally' the notion of non-relativistic conformal structure. And third: that the facts about angles are given by the distribution ensures that such facts vary smoothly from leaf to leaf of the foliation.[6]

Note also that $\mathcal{C}_{NR}$ is equivalent to $\mathcal{C}_{NR}^{\text{inf}}$, as desired. To see this, note that an equivalence class of conformally equivalent spatial metrics $[h^{\mu\nu}]$ on M is equivalent to facts about angles between intersecting worldlines on each leaf $S \subset M$ (which vary smoothly from leaf to leaf); further, one can identify $t_\mu \equiv \zeta_\mu$. Now, since ζ_μ has a component outside S (which follows from $\zeta_\mu \sigma^\mu = 0$ for all tangent vectors σ^μ to S), we can find a coordinate system such that $t_\mu = (1, 0, 0, 0)$ and for some $h^{\mu\nu} \in [h^{\mu\nu}]$, $h^{\mu\nu} = \text{diag}(0, 1, 1, 1)$, in which case the orthogonality result also follows. Conversely, given orthogonality, we clearly have that any spacelike vector σ^μ is such that $\zeta_\mu \sigma^\mu = 0$.

[5] As discussed in Chapter 1, one can also restrict one's attention to some neighbourhood $U \subset M$ and consider a 'local' notion of foliability restricted to U.

[6] Our thanks to Eleanor March for discussions here.

Thus, these 'algebraic' and 'infinitesimal' formulations of non-relativistic conformal structure are equivalent.

4.3 Building up non-relativistic structures from elementary axioms

Having now defined suitable non-relativistic versions of projective and conformal structures, we turn now to the antecedent question of whether (and, if so, how) those non-relativistic projective and conformal structures can be built up from elementary, empirically informed axioms, *per* the constructivist methodology of EPS. Roughly following the structure of §3.4, in what follows we both state the original EPS axioms and discuss how they might be modified for non-relativistic inputs. We begin with the axioms of differential topology, before turning to the light axioms, projective axioms, and compatibility axioms.

4.3.1 Differential axioms

1. *A point set $M = \{p, q \ldots\}$ is a set of events.*
2. *Light rays and particles are subsets of M.*
3. *The map $e_Q : P \to P, p \mapsto e_Q(p) = p'$ is called an echo on P from Q.*
4. *The map $m : P \to Q, p \mapsto m(p)$ is called a message from P to Q.*

Note that none of (1)–(4) are different from EPS. The difference will come from the axioms which we impose on the set of light rays; these will lead not to a relativistic conformal structure, but rather to a non-relativistic conformal structure—what we have been calling $\mathcal{C}_{NR}$.

5. *Axiom D_1: every particle is a smooth, one-dimensional manifold and any echo on P from Q is smooth and smoothly invertible.*

This axiom remains perfectly acceptable in the non-relativistic case, for there is no reason why we cannot assume smoothness in this context also.

6. *Axiom D_2: any message from a particle P to another particle Q is smooth.*

This axiom remains perfectly acceptable in the non-relativistic context.

7. *Axiom D_3: there exists a collection of triplets (U, P, P') where $U \subset M$ and $P, P' \in \mathcal{P}$ [with $\mathcal{P}$ the set of particles] such that the system of maps*

$\{x_{PP'|U}\}$ is a smooth atlas for M. Each map is written in terms of coordinates (u, v, u', v') where u and v are emission and arrival times at P and likewise on P′.

Here one will have to modify the understanding of the original EPS axiom (if not the statement *per se*), for infinitely quickly propagating light signals will not suffice to yield four *distinct* coordinates (u, v, u', v'), but rather only two, since $u = v$ and $u' = v'$. That said, the situation is not altogether desperate, for one can simply use some finitely propagating signals—drawn from the set of particles rather than the set of light rays—in order for this axiom to go through.[7]

8. *Claim: Every particle is a smooth curve in M.*

This works as long as we can construct an atlas on M, which remains true given the non-relativistic understanding of axiom D_3.

9. *Axiom D_4: Every light ray is a smooth curve in M.*

Recall that in the case of relativistic EPS, one could derive that every particle is a smooth curve in M, but the equivalent statement for light rays needed to be imposed via this axiom D_4, because one cannot 'track' the propagation of a signal travelling at the limiting speed. This remains the case in a non-relativistic version of EPS, because light signals remain the limiting speed (now infinite). Therefore, one can (and should) carry over the axiom as stated.

4.3.2 Light axioms

The two light axioms of EPS, recall, are the following:

10. *Axiom L_1: any event e has a neighbourhood V such that each event p in V can be connected within V to a particle by at most two light rays. Given such a neighbourhood and a particle P through e, there is another neighbourhood $U \subset V$ such that any event p in U can be connected with P within V by precisely two light rays and these intersect P in two distinct events e_1, e_2. If t is a coordinate function on $P \cap V$ with $t(e) = 0$, then*

[7] This is roughly akin to using the radar method for subluminal signals such as sound in discussions of 'sonic relativity'—see Todd and Menicucci (2017).

$g = -t(e_1)t(e_2)$ is a function of class C^2 on U (i.e. it is twice differentiable on U).

11. *Axiom L_2: the set L_e of light-directions at an arbitrary event e separates the projective space at e into two connected components. In the tangent space at e, the set of all non-vanishing vectors that are tangent to light rays consists of two connected components.*

In the relativistic case, EPS use these axioms to build some function g, which they then differentiate twice to obtain a conformal metric density $\mathcal{g}_{\mu\nu}$ associated with a relativistic conformal structure $\mathcal{C}$; they then show that the null cones of $\mathcal{C}$ coincide with the cones of light rays. These axioms will, however, require substantial modification—effectively, wholesale replacement—when it comes to a non-relativistic version of EPS. One can see, for example, that axiom L_1 is simply inappropriate for signals with infinite propagation speeds.

That said, an alternative approach to constructively axiomatizing a non-relativistic conformal structure does appear to be available; we propose the following replacement for axiom L_1:[8]

10'. *Axiom NR-L_1: Any neighbourhood $U \subset M$ is foliable into hypersurfaces such that any two events p and q lie on the same leaf of the foliation if and only if they are connected by light rays.*

The idea here is that light rays—being infinitely quickly propagating signals—naturally pick out, for any event $p \in U$, the events simultaneous with p. (Note that axiom NR-L_1 is essentially the first of the five axioms used by Itin and Hehl (2004) in their constructive axiomatic approach to pre-metric electromagnetism—this programme was discussed in Chapter 2, and is subjected to substantially more philosophical assessment by Chen and Read (2023).[9]) This axiom now appears to give us much of what we require for non-relativistic conformal structure, for:

- From such a foliation, one can define (cf. Chen and Read (2023)) a 'topological time' 1-form σ_μ (already discussed above), which is an arbitrary representative of a class of temporal 1-forms t_μ.

[8] To be clear: our two non-relativistic light axioms introduced below are not intended to be directly analogous to the original two relativistic light axioms of EPS.

[9] That said, the axiom of Itin and Hehl (2004) is 'global' insofar as it regards the entire manifold; we have opted for a 'local' version in terms of neighbourhoods, which should be more operationally acceptable and thus in the 'local' spirit of EPS—cf. (Chen and Read, 2023, fn. 14).

- For such a class $[t_\mu]$, there will exist classes of orthogonal $[h^{\mu\nu}]$ of spatial metrics on the leaves of the foliations.
- Together, this suffices to yield a non-relativistic conformal structure $\mathcal{C}_{\mathrm{NR}}$—albeit not yet a *unique* such structure (on which see below).

Given axiom NR-L_1, we have already managed to secure a notion of non-relativistic conformal structure. However, we have not managed to secure a *unique* conformal structure on the leaves of the foliation induced by axiom NR-L_1. We require an additional light axiom: which can be stated as follows:

11'. *Axiom NR-L_2: For any local neighbourhood U, and any $p, q, r \in S$ where $S \subset U$ is an arbitrary leaf of the foliation induced by axiom L_1, there is a determinate fact about the angles between the light rays emitted from p to q and from p to r,* mutatis mutandis *for light rays emitted from q and r.*

The point here is that angles are immediately empirically accessible—consider e.g. via the use of a sextant. Given the use of sextants here, however, it does thus seem that for non-relativistic EPS we are required to use some primitive device in addition to light rays and matter, although not a clock or a rod. In any case, since a non-relativistic conformal structure $\mathcal{C}_{\mathrm{NR}}$ just captures such facts about angles on hypersurfaces, axiom NR-L_2 thereby selects a *specific* non-relativistic conformal structure. Note that the rationale underlying axiom NR-L_2 is akin to that invoked by Barbour (2011) in his justification of shape dynamics taking spatial conformal structure as primitive,[10] writing in particular that facts about absolute angles are immediately empirically accessible. Pooley complains of Barbour's arguments here that

> [d]espite Barbour's claims, the local conformal degrees of freedom of CMC spacelike hypersurfaces are not obviously philosophically superior to the standard spacetime quantities: they are not (more) directly observable [. . .], nor are primitive temporal intervals, or primitive comparisons of distant lengths, somehow inherently suspect. (Pooley, 2013b, §6.2)

[10] It is an interesting question whether shape dynamics is therefore more operationalizable than any other 'non-relativistic' spacetime theory.

However, we are not convinced by such claims (although note again our above point regarding sextants), and there seems to us to be nothing opaque about the operationalization of angles induced by axiom NR-L_2.[11]

With all of this in hand, then, we can now proceed to move on to the projective axioms of a non-relativistic version of EPS.

4.3.3 Projective axioms

12. *Axiom P_1: Given an event e and a $\mathcal{C}$-timelike direction D at e, there exists one and only one particle P passing through e with direction D.*

We now regard any curve as being $\mathcal{C}_{\mathrm{NR}}$-timelike just in case its tangent vector is not orthogonal to any element of the equivalence class of temporal one-forms in that conformal structure. With this in mind, axiom P_1 can essentially carry over to the non-relativistic case as stated.

13. *Axiom P_2: For each event $e \in M$, there exists a coordinate system $(\overline{x}^\mu)$, defined in a neighbourhood of e and permitted by the differential structure introduced in axiom D_3, such that any particle P through e has a parameter representation $\overline{x}^\mu(\overline{u})$ with*

$$\left.\frac{d^2\overline{x}^\mu}{d\overline{u}^2}\right|_e = 0;$$

such a coordinate system is said to be projective at e.

Likewise, there is no reason why this axiom cannot carry over to the non-relativistic case. As in EPS, we want to prove that this projective structure is unique, can be measured, etc. But note that, given axioms P_1 and P_2, everything else is the same as in the relativistic case.

4.3.4 Compatibility axiom

Finally, we have to deal with the compatibility axiom. Recall that in the case of relativistic EPS, this reads:

[11] Note also that, on a non-relativistic affine structure has been constructed, axiom NR-L_2 will also at least in part operationalize the notion of spatial curvature (for the sum of all three angles from axiom NR-L_2 will be π only when space is flat—for more on which see e.g. Malament (2012, ch. 4)).

14. *Axiom C: Each event e has a neighbourhood U such that an event $p \in U$, $p \neq e$ lies on a particle P through e if and only if p is contained in the interior of the light cone v_e of e.*

For the non-relativistic version of axiom C, we propose the following:

14'. *Axiom NR-C: Each event e has a neighbourhood U such that an event $p \in U, p \neq e$ lies on a particle P through e if and only if p is not in the leaf S_e of the hypersurface foliation containing e.*

4.4 The construction of classical spacetime structures

Having presented non-relativistic notions of projective and conformal structure suitable for the constructivist methodological ambitions of EPS, and shown how such structures can be constructed via elementary, empirically informed axioms *à la* EPS, we show now that, ultimately, these structures can be used to construct a classical (i.e. non-relativistic) spacetime structure (in the sense of Malament (2012, ch. 4), via the intermediary (again analogous to the relativistic case) of a non-relativistic Weyl structure. Here, our approach is essentially to modify the results of Matveev and Scholz (2020) so as to be suitable for the non-relativistic context.

4.4.1 From non-relativistic projective and conformal structures to an NR-Weyl manifold

We begin by demonstrating, in analogy with the relativistic EPS-alternative of Matveev and Scholz (2020), (i) that $\mathcal{C}_{NR}$ and $\mathcal{P}$ are NR–EPS compatible, and (ii) that from those non-relativistic projective and conformal structures one can define the non-relativistic version of a Weyl manifold—what we'll call an 'NR-Weyl manifold'. Beginning with (i), we first offer the following definition of NR-EPS compatibility:

Definition (NR–EPS-compatibility between projective and conformal structure). *A projective structure $\mathcal{P}$ and a conformal structure $\mathcal{C}_{NR}$ are called* NR-EPS-compatible *iff the conformal spacelike geodesics of $\mathcal{C}_{NR}$ are also projective geodesics of $\mathcal{P}$ (but not necessarily vice versa).*

In order to prove that $\mathcal{C}_{\mathrm{NR}}$ and $\mathcal{P}$ as constructed in the previous section are EPS-compatible, we establish first the following lemma:

Lemma. *Each event p on S_e sufficiently close to e can be approximated arbitrarily closely by events q situated on particles through e.*

Proof. Axiom NR-C states that for any event e there is a neighbourhood such that all events within that neighbourhood *that are connected to e via particles* will not lie in S_e. If p is sufficiently close to e, then it is sufficiently close to events $q \notin S_e$ that are connected to e via particles. □

Following the lead of EPS, we need now to define a number of further non-relativistic objects. First, let $\mathscr{h}_{\mu\nu}$ be the degenerate conformal metric density associated with the equivalence class $[h^{\mu\nu}]$ given by $\mathcal{C}_{\mathrm{NR}}$, and let $\mathscr{t}_{\mu}$ be the conformal tensor density associated with the equivalence class $[t_\mu]$ given by $\mathcal{C}_{\mathrm{NR}}$. Now to the NR–EPS-compatibility proposition:

Proposition (NR–EPS-compatibility). *$\mathcal{C}_{\mathrm{NR}}$ and $\mathcal{P}$ are NR–EPS-compatible.*

Proof. Let $p \in S_e$, $p \neq e$. Let $(q_n)_{n\in\mathbb{N}}$ be a sequence of events within a $\mathcal{P}$-convex neighbourhood of e such that (i) $q_n \to p$ where '$\to$' is to denote topological convergence here, and (ii) each sequence element q_n together with e defines the sequence element P_n in the series of geodesics (particles) $(P_n)_{n\in\mathbb{N}}$. (Note that (i) is a legitimate demand in light of the previous lemma (based itself on Axiom *NR–C*).) Notably, each particle P_n can be taken to obey the (projective) geodesic equation $\ddot{x} + \Pi^{\lambda}_{\mu\nu}\dot{x}^{\mu}\dot{x}^{\nu} = \lambda_n \dot{x}$ parameterized in such a way that, for $\lambda_n = 0$, P_n goes through e, and, for $\lambda_n = 1$, P_n goes through q_n. In particular, in the limit $q_n \to p$, the corresponding geodesic $P \leftarrow P_n$ goes through e, and for $\lambda = 1$ the geodesic goes through $p \in S_e$; and the sequence $\{T_n\}_{n\in\mathbb{N}}$ where T_n is the tangent vector of P_n at e converges to the tangent vector T of that geodesic P. (All of this assumes that q_n, and p lie in a sufficiently small $\mathcal{P}$-convex coordinate neighbourhood of e.)

For $T_n \to T$ with T_n timelike, T is either timelike or spacelike: for any tangent vector s, the map $s^\mu \mapsto \mathscr{t}_\mu s^\mu$ is a continuous function, i.e. convergence of a sequence s^μ_n, implies the convergence of $\mathscr{t}_\mu s^\mu$. As for all T_n, $\mathscr{t}_\mu T^\mu_n > 0$, thereby $\mathscr{t}_\mu T^\mu \geq 0$. But we can exclude that T^μ is timelike: for if it were timelike, then P—coming from e—would have to intersect S_e a second time—this time as a past-directed vector, which is against the (implicit) assumption that P_n (and thus also P) are everywhere future-directed.

We now want to show that, between passing e and p, P can only contain events on S_e, i.e. $P \subset S_e$. Assume $q \in P$ was a point strictly to the future of S_e upon passing from e to p. Then: $q \in S_q$ with $S_q \cap S_e = \emptyset$. Then a sufficiently close p could not be reached by P anymore. Thus, q must be on S_e.

To summarize: (1) the tangents along P are spacelike between e and p (what we said for the tangent vector at p holds without any restriction for the tangents at the points to P between e and p too). P is thereby a projective spacelike geodesic. (2) Between e and p, P is contained in S_e. With an additional assumption (and only then)—see discussion below—it follows then that P is a $\mathcal{C}_{\text{NR}}$-spacelike geodesic.

Given that being a geodesic is a local property and that the considerations above hold around any other event e, it follows that for every spacelike geodesic a projective spacelike geodesic can be constructed that coincides with it. □

Note here that step (2) is warranted only if one assumes that the propagating physical signals on any S_e can only move along conformal geodesics—this makes sense insofar as one thinks that the only signals which can propagate on such surfaces are light rays. Clearly, this is a substantial physical assumption, and a substantial difference between the relativistic and non-relativistic versions of EPS (in the former case, no such assumption was required, because one could prove that ν_e is a $\mathcal{C}$-null cone). (One way to appreciate the difference here is by realizing that null surfaces are just fundamentally distinct from spacelike hypersurfaces: given the generating null vector field, curves deviating from the integrating curves cannot be null (Galloway, 2004, §3).)

We turn next to (ii), which will require a few more definitions:

Definition (NR–Weyl space). *A manifold M together with a NR–EPS-compatible pair of non-relativistic conformal structure $\mathcal{C}_{NR}$ and projective structure $\mathcal{P}$ is called a* NR–Weyl space $(M, \mathcal{C}_{NR}, \mathcal{P})$.

Of course, a choice of $\mathcal{C}_{\text{NR}}$ determines a class $[h^{\mu\nu}]$ of degenerate spatial metrics, and a class $[t_\mu]$ of clock forms orthogonal to those spatial metrics. Given this, we can then define:

Definition (NR–Weyl structure and NR–Weyl manifold).

1. *A (pseudo-Riemannian)* NR–Weyl structure *is given by the triple* $(M, \mathcal{C}_{NR}, \nabla)$ *where M is a differentiable manifold;* $\mathcal{C}_{NR}$, per *the above,*

fixes classes $[h^{\mu\nu}]$ *and orthogonal* $[t_\mu]$*, and* ∇ *is the covariant derivative of a torsion-free affine connection* Γ *(called* NR–Weyl connection*), constrained by the* NR–Weyl compatibility conditions *that for any pair of* t_μ *and* $h^{\mu\nu}$ *in their respective equivalence classes, there is a differential 1-form* φ_μ *such that*

$$\nabla_\lambda h^{\mu\nu} = -\varphi_\lambda h^{\mu\nu}, \tag{4.1}$$

$$\nabla_\lambda t_\mu = \frac{1}{2}\varphi_\lambda t_\mu. \tag{4.2}$$

(Cf. e.g. Wolf et al. (2023a, p. 21).)

2. *A* NR–Weyl manifold $(M, [(t_\mu, h^{\mu\nu}, \varphi_\mu)])$ *is a differentiable manifold* M *endowed with a* NR–EPS-Weyl metric *defined by an equivalence class of triples* $(t_\mu, h^{\mu\nu}, \varphi_\mu)$*, where* t_μ *is a degenerate temporal 'metric' and* $h^{\mu\nu}$ *a degenerate spatial 'metric' on* M *and* φ_μ *is a (real valued) differential 1-form on* M*. Equivalence is defined by conformal rescalings* $t_\mu \mapsto \tilde{t}_\mu = \Omega^{-1/2} t_\mu$, $h^{\mu\nu} \mapsto \tilde{h}^{\mu\nu} = \Omega h^{\mu\nu}$ *and the corresponding transformation* $\varphi_\mu \mapsto \varphi_\mu - d_\mu \ln \Omega$*.*[12]

As in the relativistic case, both structures—i.e. that of a NR-Weyl structure and that of a NR-Weyl manifold—are straightforwardly provably equivalent. To see this, note that (4.1) and (4.2) are invariant under the transformations $t_\mu \mapsto \tilde{t}_\mu = \Omega^{-1/2} t_\mu$, $h^{\mu\nu} \mapsto \tilde{h}^{\mu\nu} = \Omega h^{\mu\nu}$ and $\varphi_\mu \mapsto \varphi_\mu - d_\mu \ln \Omega$, and thereby pick out equivalence classes $(t_\mu, h^{\mu\nu}, \varphi_\mu)$ related by said transformations and so a NR-Weyl manifold. On the other hand, derivative operators compatible with the t_μ and $h^{\mu\nu}$ in the conformal equivalence classes in an NR-Weyl manifold will (when acting on other elements of those equivalence classes) satisfy (4.1) and (4.2), and thereby suffice to define an NR-Weyl structure.

Analogously to the relativistic case, the family of φ_μ defining a Weyl structure/Weyl manifold differ only by exact forms; let us call the common exterior derivative NR-distant curvature (*Streckenkrümmung*), denoted by $f_{\mu\nu}$. That is, $f_{\mu\nu}$ is given by $f_{\mu\nu} = d_\mu \varphi_\nu = d_\mu \tilde{\varphi}_\nu = \ldots$.[13]

Now we recall that in the context of a classical spacetime, specifying a timelike vector field v^μ, one can associate with said vector field a 'special' compatible

[12] These transformations are not the same as those discussed by Dewar and Read (2020), for the reasons given in footnote 9 of that article.

[13] As before, the distant curvature is only well-motivated along these lines if exact forms are closed. A sufficient condition which will be required from now on for neighbourhoods is simply-connectedness.

connection, with components given by Bekaert and Morand (2016, eq. 3.11)

$$\overset{\nu}{\Gamma}{}^{\lambda}_{\mu\nu} = \nu^{\lambda}\partial_{(\mu}t_{\nu)} + \frac{1}{2}h^{\lambda\rho}\left(\partial_{\mu}\overset{\nu}{h}_{\rho\nu} + \partial_{\nu}\overset{\nu}{h}_{\rho\mu} - \partial_{\rho}\overset{\nu}{h}_{\mu\nu}\right). \tag{4.3}$$

Unlike in the relativistic case, the inverse of $h^{\mu\nu}$ is not unique, but instead can only be specified relative to a choice of ν^{μ}, *per* (Bekaert and Morand, 2016, eq. 2.9)

$$\overset{\nu}{h}_{\mu\nu}\nu^{\nu} = 0, \tag{4.4}$$

$$\overset{\nu}{h}_{\lambda\nu}h^{\mu\lambda} = \delta^{\mu}_{\nu} - \nu^{\mu}t_{\nu}; \tag{4.5}$$

It is this 'inverse spatial metric' $\overset{\nu}{h}_{\mu\nu}$ which appears in (4.3). Now, for any other choice of timelike vector field for which a connection is not special, the components of that connection can be written as (Bekaert and Morand, 2016, eq. 3.12)

$$\Gamma^{\lambda}_{\mu\nu} = \overset{\nu}{\Gamma}{}^{\lambda}_{\mu\nu} + h^{\lambda\rho}t_{(\mu}\overset{\nu}{F}_{\nu)\rho}, \tag{4.6}$$

where $\overset{\nu}{F}_{\mu\nu}$ is known as the 'Newton–Coriolis 2-form' relative to ν^{μ}, which has components (Bekaert and Morand, 2016, p. 17)

$$\overset{\nu}{F}_{\alpha\beta} := -2\overset{\nu}{h}_{\lambda[\alpha}\nabla_{\beta]}\nu^{\lambda}. \tag{4.7}$$

All of the above is standard in the contemporary physics literature on non-relativistic gravity—for some recent philosophical presentation and assessment, see Teh (2018). Note that the specific choice of timelike vector field ν^{μ} can be regarded as being an additional conventional input required to fix a classical spacetime structure—one not required in the relativistic physics, where the metric $g_{\mu\nu}$ has a unique associated Levi-Civita connection. In what follows, we will indeed treat this choice of ν^{μ} as a conventional one; we leave open the possibility that it could be operationalized in some way or other (thereby vitiating the need for this object to be fixed by convention), perhaps by appealing (as Ewen and Schmidt (1989) do) to the trajectories of 'gravitating matter'.[14]

[14] One reason why we—*qua* theory reductionists—would expect the choice of ν^{μ} to be conventional is that its value cannot be derived from GR; rather, the freedom associated with this field emerges in the non-relativistic limit: see Hartong et al. (2022).

What one would like to do now is to use (4.3) as a guide to writing down the components of a non-relativistic special conformal connection, by analogy with (1.3). In order for (4.5) to be invariant under the conformal transformations of $h^{\mu\nu}$ and t_μ associated with an NR-Weyl manifold (see above), one requires that

$$v^\mu \mapsto \Omega^{1/2} v^\mu, \tag{4.8}$$

$$\overset{v}{h}_{\mu\nu} \mapsto \Omega^{-1} \overset{v}{h}_{\mu\nu}. \tag{4.9}$$

Associated with the ensuing equivalence classes $[v^\mu]$ and $[\overset{v}{h}_{\mu\nu}]$, one can then define the conformal tensor densities $\boldsymbol{v}^\mu$ and $\overset{\boldsymbol{v}}{\boldsymbol{h}}_{\mu\nu}$, respectively; in turn, then, one can define the components of a non-relativistic special conformal connection as follows.

Definition. *Relative to a $\boldsymbol{v}^\mu$, the components of the non-relativistic special conformal connection are given by*

$$K^\lambda_{\mu\nu} := \boldsymbol{v}^\lambda \partial_{(\mu} \boldsymbol{t}_{\nu)} + \frac{1}{2} \boldsymbol{h}^{\lambda\rho} \left(\partial_\mu \overset{\boldsymbol{v}}{\boldsymbol{h}}_{\rho\nu} + \partial_\nu \overset{\boldsymbol{v}}{\boldsymbol{h}}_{\rho\mu} - \partial_\rho \overset{\boldsymbol{v}}{\boldsymbol{h}}_{\mu\nu} \right).$$

Clearly, by construction this object is invariant under the non-relativistic conformal transformations associated with an NR-Weyl manifold. Now, just as in the relativistic case, we have the following.

Definition. *The difference between the projective and conformal connection is given by*

$$\Delta^\mu_{\nu\lambda} := \Pi^\mu_{\nu\lambda} - K^\mu_{\nu\lambda}. \tag{4.10}$$

$\Delta^\mu_{\nu\lambda}$ can now be shown to fulfil certain conditions. First of all:

Lemma (Properties of $\Delta^\mu_{\nu\lambda}$). *$\Delta^\mu_{\nu\lambda}$ satisfies $\Delta^\mu_{[\nu\lambda]} = \Delta^\mu_{\nu\mu} = 0$.*

Proof. For showing $\Delta^\mu_{[\nu\lambda]} = 0$, note that we have already seen that $\Pi^\mu_{[\nu\lambda]} = 0$, so it suffices to show that $K^\mu_{[\nu\lambda]} = 0$. The proof for this is exactly analogous to that for the relativistic case.

For $\Delta^{\mu}_{\nu\mu} = 0$, consider

$$K^{\nu}_{\mu\nu} = \mathscr{v}^{\nu}\partial_{(\mu}\mathscr{t}_{\nu)} + \frac{1}{2}\mathscr{h}^{\nu\rho}\left(\partial_{\mu}\overset{\mathscr{v}}{\mathscr{h}}_{\rho\nu} + \partial_{\nu}\overset{\mathscr{v}}{\mathscr{h}}_{\rho\mu} - \partial_{\rho}\overset{\mathscr{v}}{\mathscr{h}}_{\mu\nu}\right) = 0;$$

where the ultimate and penultimate terms cancel immediately, and the first two terms by using (4.5). Since we have already argued above that for a projective connection one can take $\Pi^{\mu}_{\nu\mu} = 0$, the result follows. □

With all of the above definitions in hand, we can turn now—by analogy with the constructions presented on behalf of EPS by Matveev and Scholz (2020)—to algebraic conditions for the non-relativistic versions of EPS compatibility.

Corollary (Algebraic condition for NR-EPS-compatibility). *The following condition holds:*

$$\Delta^{\mu}_{\nu\lambda} = -\eta_{(\nu}\delta^{\mu}{}_{\lambda)} + \phi^{\mu}t_{\nu}\varphi_{\lambda}. \tag{4.11}$$

where η_{μ} and φ_{μ} are 1-forms and ϕ^{μ} is a vector.

(Note that, for a specific choice of φ_{μ}, this is projectively equivalent to the difference tensor for non-relativistic Weylian theories presented at Wolf et al. (2024a, p. 20), which was derived by taking the non-relativistic limit of a relativistic non-metric theory.) To prove this corollary, we will modify the proof of the analogous algebraic condition for EPS-compatibility given in the relativistic case by Matveev and Scholz (2020):

Proof. For any spacelike geodesic γ, by analogy with the relativistic case, we have

$$\nabla^{\mathscr{h}}_{\dot{\gamma}}\dot{\gamma} = 0 \tag{4.12}$$

$$\nabla^{\Gamma}_{\dot{\gamma}}\dot{\gamma} = \beta(\gamma, \dot{\gamma})\,\dot{\gamma}, \tag{4.13}$$

for some function β, and where $\mathscr{h}^{\mu\nu}$ is the degenerate conformal metric density associated with the equivalence class $[h^{\mu\nu}]$ given by $\mathcal{C}_M$.[15] Then subtracting gives

[15] As in the relativistic case, one could follow the lead of Matveev and Scholz (2020) and use representatives of $[h^{\mu\nu}]$ rather than the 'invariant' tensor density; if one proceeds along this route, one can argue that if the first condition obtains for some $h \in \mathcal{C}_M$, then it holds for all $h \in \mathcal{C}_M$.

$$\underbrace{(\Pi^{\mu}_{\nu\lambda} - K^{\mu}_{\nu\lambda})\dot{\gamma}^{\lambda}}_{=\Delta^{\mu}_{\nu\lambda}\dot{\gamma}^{\nu}} = \beta(\gamma,\dot{\gamma})\dot{\gamma}^{\mu}. \tag{4.14}$$

where, just as in the relativistic case, $\Pi^{\mu}_{\nu\lambda}$ are the components of the projective connection under consideration, and $K^{\mu}_{\nu\lambda}$ are the components of the non-relativistic conformal connection under consideration.

For any vector v^{μ}, consider now the order-three polynomial in v^{μ}

$$\Delta^{\mu}_{\nu\lambda}v^{\nu}v^{\lambda}v^{\rho} - \Delta^{\rho}_{\nu\lambda}v^{\nu}v^{\lambda}v^{\mu}. \tag{4.15}$$

We see immediately that this expression must be zero for v^{μ} spacelike (as then $\ldots = \beta v^{\mu}v^{\beta} - \beta v^{\beta}v^{\mu} = 0$). So, the zero points for the order-three polynomial must include spacelike vectors v^{μ}, i.e. vectors for which $t_{\mu}v^{\mu} = 0$. Given that multivariate polynomials can be uniquely decomposed into irreducible factors—and $t_{\nu}v^{\nu}$ is such a factor—the order-three polynomial decomposes into $(t_{\nu}v^{\nu})\omega^{\mu\rho}_{\tau\sigma}v^{\tau}v^{\sigma}$. Again, the contravariant 2-form $\omega^{\mu\rho}_{\tau\sigma}v^{\tau}v^{\sigma}$ has rank two such that it is given by $(\phi^{\mu}v^{\rho} - \phi^{\rho}v^{\mu})(\varphi_{\lambda}v^{\lambda})$ where the factor $(\varphi_{\lambda}v^{\lambda})$ (with φ some vector) comes in from the requirement that the polynomial is exactly of order three. This gives rise to the equation $\Delta^{\mu}_{\nu\lambda}v^{\nu}v^{\lambda}v^{\rho} - \Delta^{\rho}_{\nu\lambda}v^{\nu}v^{\lambda}v^{\mu} = (\phi^{\mu}v^{\rho} - \phi^{\rho}v^{\mu})(t_{\nu}v^{\nu})(\varphi_{\lambda}v^{\lambda})$. In analogy to the relativistic case, the solution is again the general solution to the homogeneous system, plus one specific solution, which in this case means $\Delta^{\mu}_{\nu\lambda} = \delta^{\mu}_{\nu}\eta_{\lambda} + \delta^{\mu}_{\lambda}\eta_{\nu}$ for the general solution, and $\Delta^{\mu}_{\nu\lambda} = -\phi^{\mu}t_{\nu}\varphi_{\lambda}$ for a specific solution, resulting in the solution $\delta^{\mu}_{\nu}\eta_{\lambda} + \delta^{\mu}_{\lambda}\eta_{\nu} - \phi^{\mu}t_{\nu}\varphi_{\lambda}$. □

Note also that, just as the relativistic condition for EPS-compatibility was identified in a very different context by Wolf et al. (2023a, eq. 55), so too was a particular form of the non-relativistic condition for NR–EPS compatibility identified by Wolf et al. (2024a, eq. 58).

Claim (Existence of a NR-Weyl connection). *The expression*

$$\Gamma^{\mu}_{\nu\lambda} = K^{\mu}_{\nu\lambda} - \phi_{(\nu}\delta^{\mu}{}_{\lambda)} + \phi^{\mu}t_{\nu}\varphi_{\lambda} \tag{4.16}$$

defines a NR-Weyl connection on the NR–EPS–Weyl space $(M, \mathcal{C}_{NR}, \mathcal{P})$ *where* $\phi_{\lambda}, q^{\mu} = h^{\mu\lambda}\phi_{\lambda}$.

Proof. Just as in the relativistic case, this follows from the corollary just before by using a representation of the projective structure in which η and ϕ are the same. That Γ is a NR-Weyl connection is easily checked from its transformation properties. □

To summarize: analogously to the relativistic case, the logic here was as follows:

1. Show that $\mathcal{C}_{NR}$ and $\mathcal{P}$ are NR-EPS-compatible.
2. Specify an algebraic condition for NR-EPS compatibility.
3. Demonstrate that if said algebraic condition is satisfied, then an NR-Weyl connection is thereby determined.

All three of these tasks have now been achieved.

4.4.2 From NR-Weyl manifolds to classical spacetimes

Again following the lead of EPS in the relativistic case, we now need to provide a characterization of a classical spacetime manifold in terms of an NR-Weyl manifold. In the relativistic case, the analogous gap (between a Weyl spacetime and a Lorentzian spacetime) was bridged by way of two equivalent criteria: the 'equidistance criterion' and the 'congruence criterion'. Recall that the former of these reads as follows—a statement just as adequate to the non-relativistic case as to the relativistic case:

> **Equidistance criterion** the proper times t, t' on two arbitrary, infinitesimally close, freely falling particles P, P' are linearly related (to first order in the distance) by Einstein-simultaneity; i.e., whenever $p_1, p_2, \ldots$ is an equidistant sequence of events on P (ticking of a clock) and $q_1, q_2, ..$ is the sequence of events on P' that are Einstein-simultaneous with $p_1, p_2, \ldots$ respectively, then $q_1, q_2, \ldots$ is (approximately) an equidistant sequence on $P' \ldots$ (p. 69)

Before we proceed further, three comments are in order. First: we have already seen that axiom P_2—unmodified between the relativistic and non-relativistic cases—underwrites the torsion-freeness of spacetime (recall our discussion of this point from Chapter 1). Now, in non-relativistic spacetimes the condition $d_\mu t_\nu = 0$ is both necessary and sufficient for the absence of twins paradox-type effects ('first-clock effects'), and for a global notion of proper time (which amounts to the equidistance criterion) (Bekaert and Morand, 2016). But from vanishing spacetime torsion $d_\mu t_\nu = 0$ follows (see again e.g. Bekaert and Morand (2016)), so in the non-relativistic case, the equidistance criterion falls directly out of axiom P_2, thereby setting it on (one might argue) more solid empirical footing.

Second: one might be surprised to see Einstein-simultaneity (i.e. simultaneity *per* a clock-and-mirrors radar construction) appearing in the discussion of non-relativistic physics. Of course, this is somewhat odd, insofar as there is no ambiguity about the simultaneity of distant clocks given the availability of infinitely quickly propagating signals with which to synchronize them. Nevertheless, the notion of Einstein-simultaneity is still perfectly comprehensible and meaningful in a non-relativistic context.

Our third comment about the equidistance criterion is this: setting $d_\mu t_\nu = 0$ imposes only that temporal non-metricity be symmetric; it doesn't impose that the non-metricity 1-form φ_μ be integrable (as in the relativistic case). So, although in some sense the equidistance criterion is better grounded empirically in the non-relativistic case (see our first comment), it is also not useful for instigating the move from an NR-Weyl structure to a classical spacetime structure. With this in mind, then, we focus in what follows on the congruence criterion. In the non-relativistic case, this can be stated as follows:

> **Congruence criterion:** Consider parallel transport of a vector V_p from a point p to a point q along two different curves P, P'. The resulting vectors, at q, V_q and V'_q, will in general be different. If and only if V_q and V'_q are congruent (i.e. have the same length) for all such figures will it be the case that the NR-Weyl geometry considered is in fact metric and non-relativistic (a 'classical spacetime', in the sense of (Malament, 2012, ch. 4)).

With this statement of the congruence criterion in hand, we begin by adapting the following lemma from the relativistic case:

Lemma. *(Congruence criterion for classical spacetime geometry) In any simply connected domain of M, there is a spatial metric $h^{\mu\nu}$ and orthogonal clock form t_μ, both compatible with the conformal structure $\mathfrak{C}_{NR}$, such that there is a NR-Weyl connection which is metric with respect to $h^{\mu\nu}$ and t_μ iff the distant curvature $f_{\mu\nu} = 0$.*

(Note: the reader should not confuse here the distant curvature with the Newton–Coriolis 2-form.)

Proof. *Left-to-right*: If Γ is metric, then by definition $\nabla_\mu h^{\nu\lambda} = \nabla_\mu t_\nu = 0$, in which case the non-metricity 1-form $\varphi_\mu = 0$, and so $f = 0$.

Right-to-left: If $f = 0$, then $d\varphi = 0$, in which case locally $\varphi = d\chi$ for some scalar field χ. But in that case, one is working in an integrable NR-Weyl geometry; and in such a case, one can perform a NR-Weyl gauge transformation (i.e. transformations on $(h^{\mu\nu}, t_\nu, \varphi)$ which fixes the same NR-Weyl geometry)

$$t_\mu \mapsto \tilde{t}_\mu := \Omega^{-1/2} t_\mu,$$
$$h^{\mu\nu} \mapsto \tilde{h}^{\mu\nu} := \Omega h^{\mu\nu},$$
$$\varphi_\mu \mapsto \tilde{\varphi}_\mu := \varphi_\mu - d_\mu \ln \Omega.$$

in order to transform to a gauge (the equivalent of the 'Riemann gauge' in the relativistic case) in which $\tilde{\varphi} = 0$. One then finds that the metricity condition is satisfied. □

More rigorously then, the congruence criterion can be established as follows:

Proposition. *(Congruence criterion for classical spacetime geometry) In any simply connected domain of M, there is a spatial metric $h^{\mu\nu}$ and orthogonal clock form t_μ, both compatible with the conformal structure $\mathcal{C}_{NR}$, such that the NR-Weyl connection Γ is metric with respect to $h^{\mu\nu}$ and t_μ iff there is no second clock effect.*

Proof. The second clock effect in a NR-Weyl geometry occurs iff there is non-vanishing length curvature $f_{\mu\nu}$. This can be seen by explicitly comparing the result of parallel transporting the same starting vector via two different paths from a point p to a point q: there will only be no change in length iff φ_μ (as defined in the previous proof) is closed, i.e. $d_\mu \varphi_\nu = 0$. But this is again directly equivalent to $f_{\mu\nu} = 0$. □

4.5 Philosophical discussion

In this section, we'll now consider some of the philosophical upshots of the non-relativistic version of the EPS axiomatization which has been offered in this chapter. Specifically, we'll consider how our axiomatization relates to non-relativistic versions of Weyl's theorem (§4.5.1), how the non-relativistic approach relates to the original version of EPS (§4.5.2), the limitations of our non-relativistic axiomatization (§4.5.3—in analogy with our discussion closing Chapter 1), and issues of conventionalism in the non-relativistic approach (§4.5.4).

4.5.1 Non-relativistic versions of Weyl's theorem

Recall from our discussion in Chapter 1 that Weyl himself is best understood as offering a uniqueness result, in what has come to be known as *Weyl's theorem*: for a given Weyl geometry, that geometry is fixed uniquely by its associated projective and conformal structures. Malament (2012) presents a special case of this result, when he demonstrates that a Lorentzian spacetime geometry (this, of course, being a special case of a Weylian spacetime geometry, in which metric compatibility has been enforced) is likewise fixed uniquely (i.e. up to a global constant) by its associated projective and relativistic conformal structures. On the other hand, what EPS themselves offer, *inter alia*, is an existence result: for any given piece of projective structure and any given piece of relativistic conformal structure (Weyl-compatible with that projective structure), one can construct from those two structures a Weyl spacetime geometry.

It is worth clarifying the relationships between the analogous theorems which have now been proven in the non-relativistic case. In fact, the one remaining piece of the puzzle here—i.e. the theorem which has yet to be proven—is a non-relativistic version of Weyl's theorem itself, which would state that a NR-Weyl geometry is fixed uniquely (up to a constant) by its projective and conformal structures. Such a theorem is in fact true; the proof is exactly analogous to that sketched for the relativistic case in Chapter 2.

With this in mind, recall e.g. the theorem proved by Curiel (2015) (one could likewise call to mind the theorem proved by Ewen and Schmidt (1989), although one should remember that the notion of non-relativistic conformal structure deployed in that latter case is different from that used by Curiel (2015) or by March (2023)). This states that a Newton-Cartan spacetime geometry is fixed uniquely (up to a constant) by its associated projective and non-relativistic conformal structures. In parallel with the relationship between the theorem proved by Weyl versus the theorem proved by Malament, one now likewise sees here that Curiel's theorem is a special case of the non-relativistic 'Weylian' result, in which a non-metricity 1-form is generically permitted. And naturally, both of these results qualify as 'uniqueness' results, as compared with the non-relativistic, EPS-inspired 'existence' result which has been sketched above.

4.5.2 Relation to relativistic EPS

It is by now well understood that there are many ways in which one can take the Newtonian limit of relativistic physics—*inter alia*, (i) a geometrical approach via 'widening' the light cones (loosely corresponding to taking $c \to \infty$—but more rigorously understood in terms of the 'frame theory' of Ehlers (2019), on which see (Fletcher, 2019) for recent philosophical discussion); (ii) 'null reduction'—i.e. via a Kaluza–Klein-style compactification/projection of dynamics on some higher-dimensional space (see e.g. (Bagchi et al., 2022); the approach was recently deployed by philosophers in (Read and Teh, 2018)); (iii) via a $1/c$ expansion of relativistic objects (see e.g. (Dautcourt, 1997)). Focusing here on approach (i) (on which see also Malament (1986)), as one widens the lightcones of a relativistic spacetime structure one can show that

$$g_{\mu\nu} \to t_\mu t_\nu \tag{4.17}$$

$$g^{\mu\nu} \to h^{\mu\nu}. \tag{4.18}$$

With this limit in hand, it is almost immediate to see that the non-relativistic limit of a relativistic conformal structure $\mathcal{C}$ *just is* a non-relativistic conformal structure $\mathcal{C}_{\mathrm{NR}}$ (for further discussion on this point, see March (2023)); moreover, the output of the non-relativistic version of EPS is a classical spacetime structure, which *just is* the non-relativistic limit of the output of EPS—i.e. a Lorentizan spacetime structure. Thus, in a sense, we have completed a commuting diagram: the non-relativistic limits of the EPS inputs yield a non-relativistic version of conformal structure, which in turn (with projective structure) yields the non-relativistic limit of a relativistic spacetime structure. (The above, of course, is nearly correct, but one must not forget that, in the non-relativistic limit, one must *also* input a congruence of timelike vector fields representing gravitational matter in order to recover a *specific* 'special' non-relativistic compatible connection!)

4.5.3 Limitations of non-relativistic EPS

Many of the limitations of the relativistic version of EPS discussed in Chapter 1 carry over to the case of a non-relativistic version of the EPS methodology which has been outlined in this chapter. For example, there are still concerns about locality restrictions, and the use of idealizations, and non-constructivist deviations from the ideal of semantic linearity at various points.

Let us focus here on one particular issue which was also raised in Chapter 1 with regard to EPS: the fact that the construction is essentially *kinematical*, rather than dynamical. As such, the approach yields only a classical spacetime structure, but it does not yield the specific dynamics of e.g. Newton-Cartan theory as presented in its modern form by e.g. Malament (2012).

This being said, perhaps there is room to operationalize more of the dynamical content of Newton-Cartan theory than one might initially think. Recall from above that axiom NR-L_2 affords at least partial operational access to the spatial curvature of a (to-be constructed) non-relativistic spacetime. But now recall from Malament (2012, §4.3) that spatial flatness together with the admission of a timelike vector field which is rigid and twist-free is equivalent to satisfaction of the Newton–Poisson equation of standard Newtonian gravity (in turn equivalent, via Trautman geometrization, to satisfaction of the geometrized Poisson equation of Newton-Cartan theory). Therefore, if the latter condition could also be operationalized—and, indeed, we see no in-principle reason why this could not be the case—then one would thereby come to have operationalized at least some of the *dynamical* content of Newtonian gravity: surely an interesting result.

Note also that the fact that one does not obtain the *specific* dynamics of e.g. Newton-Cartan theory could be construed as an advantage of the approach, for it renders it consistent with a richer class of non-relativistic theories considered by e.g. Hartong et al. (2022). That said, many of these spacetimes are torsionful (the so-called 'twistless-torsionful Newton-Cartan' paradigm), in which case they would appear to be ruled out by axiom P_2. Given that these theories are often motivated on the grounds that they can be obtained from a more sophisticated ('type II') limit of general relativity, this arguably constitutes some motivation to drop axiom P_2.

4.5.4 Conventionalism in non-relativistic EPS

To what extent can a given classical spacetime be recovered via the existence of a local directing field, in the manner of Coleman and Korté, discussed in Chapter 2? On this question, Coleman and Schmidt write the following:

> The second case concerns the geometrization of nonrelativistic gravity. The motion of neutral monopoles is governed by a (Cartan) connection in this case as well and the Newtonian theory of gravitation can be formulated without imposing an additional flat connection. [. . .] It is known that the

> structure of Newton-Cartan space-time is uniquely determined by its projective structure, the simultaneity relation, and the motion of gravitating matter.
>
> In neither of these cases, however, has an analog of the constructive axiomatics for GTR yet been worked out in which the projective criterion plays a suitable role. Indeed, in the nonrelativistic cases the projective criterion, either the Pasch criterion presented above or the cubic criterion, cannot by itself serve to single out a unique directing-field structure because a great many if not all turn out to be projective. As Coleman and Korté have pointed out, it is only in the relativistic case that there exists a unique geodesic directing field corresponding to a unique projective structure that determines the free-fall motions of neutral massive monopoles. (Coleman and Schmidt, 1995, p. 1344)

(Here, the reference is to (Coleman and Korté, 1994).) The point here seems to be that, in the non-relativistic case, one can always find a new geodesic directing field by exploiting the Milne symmetry of the theory (on which see e.g. (Bekaert and Morand, 2016))—essentially, the freedom to choose different v^{μ}. As such, this would appear to make the case for geometric conventionalism *stronger* in the non-relativistic context than in the relativistic context—corroborating (albeit for very different reasons, of course) recent arguments to this effect made by Weatherall and Manchak (2014) and (separately) March et al. (2024). Note, though (as already pointed out above), that if one were able to operationalize a specific choice of v^{μ}, then this further concern about conventionalism would not arise.

Conclusions

After 100 years of Reichenbachian constructive axiomatics, and 50 years of EPS, we hope that this book will serve as strong evidence that there remains a great deal of interest in, and potential for, the constructive axiomatic method. Over the course of the previous four chapters, we have tried to pay tribute to our heroes Reichenbach, Weyl, and Ehlers–Pirani–Schild, by giving both EPS and constructive axiomatics at large their long-missing comprehensive appraisal. With this book, the reader has been offered a pedagogical presentation of EPS, a contexualization of it, and arguments for and against various of its readings; but hopefully more than that: through our two application cases the reader has seen how constructive axiomatization can be put to action to do real work.

Let us go through each of these achievements in slightly more detail:

Pedagogical presentation of EPS: EPS is difficult to parse; the surrounding literature is scattered. We have provided a pedagogical presentation (including all relevant proofs) and linked up the original EPS axiomatization to follow-up works.

Contextualization: Much of what we said in the contextualization, first of all, was aimed at helping to better understand what EPS is: where it comes from, with which major problems it has to contend, what variants there are; and we also linked this type of axiomatization to many related themes from philosophy (constructive mathematics, theory formulation, empirical interpretation, operationalism, constructive empiricism, among others).

Readings of EPS: An important aim of this book has been to explore the various options on how to read EPS; as a result of our efforts; in particular, even those sceptical of EPS for a seemingly naïve foundationalist programme should find some readings to render the EPS axiomatization worthwhile. Among all readings, we maintain that the reading of EPS as a non-exhaustive account of the empirical meaning of theoretical terms to an agent is not just robust but very attractive.

Application: The EPS approach has interesting potential applications to open scientific problems. By applying the EPS approach to

Constructive Axiomatics for Spacetime Physics. Emily Adlam, Niels Linnemann, and James Read, Oxford University Press. DOI: 10.1093/9780198922391.003.0006

non-relativistic spacetime theory, we understand the operational character of classical spacetimes much better than previously ever achieved; notions of non-relativistic conformal structure in turn have unexpected applications to e.g. non-relativistic twistor theory (on which see March (2023)). And applying EPS in a quantum context helps to highlight a number of issues surrounding the appropriate formulation of a quantum theory of gravity. It is interesting that the problem of point identification across branches takes on so much significance in the formulation of a quantum EPS kinematics, and it remains to be seen whether this problem should be resolved at the level of the dynamics or the kinematics. As we have noted, a resolution of this question seems important not only for quantum gravity but also for a full understanding of the fate of the hole argument in quantum gravitational spacetimes.

Related to this: after having studied constructivist approaches such as EPS, one accrues better facility with sub-metrical constituents such as projective and conformal structures (whether relativistic or non-relativistic), and how those structures might be applied in future physical theorizing (cf. (Stachel, 2011)). That is, one obtains better facility with the tools of the foundational trade.

At the same time, we thus learn a lot about the axiomatization of GR through these applications: the axiomatization of GR can have a strong operational character because light is very explicitly modelled in GR (in NCT, it is modelled in terms of (conformal) spacelike null geodesics but that makes the trajectories of light less distinguishable from others). Secondly, the axiomatization of GR can have a strong constructive character because it is a classical theory (in quantum gravity, we need to give up the immediacy as we can usually only access one branch at a time).

There are various other topics which we have not explored in this book and which remain open for further investigation. To mention here just three:

Formal nature of the axiomatization: We expect interesting lessons about the inner workings of EPS (and constructive axiomatisation more generally) from formally modelling the logical structure of the axiomatisation system set up by EPS, including a more rigorous distinction between forms of definitions and axioms, respectively.

Application: We are curious about the extent to which constructive-constructivist axiomatization are possible outside (1) the spacetime and

(2) the classical context. Further case studies (such as on quantum mechanical theories, perhaps in a transition amplitude, process-oriented formulation *à la* Barandes (2023a, 2023b)) offer themselves here.

Quantum gravity: In Chapter 3, we made a preliminary investigation into the possibility of an EPS-style approach in the context of quantum mechanics, but many more ideas could be explored in this vicinity. In particular, a number of existing theories of quantum gravity encounter challenges when it comes to the issue of recovering classical spacetime in an appropriate limit, so is possible that connecting these approaches up with classically defined operational axioms in the style of EPS might offer new insight into this problem. For example, perhaps one way to address the problem of time would be to make use of an operational axiomatization of the notion of a clock (possibly akin to some of the clock constructions considered in Chapter 2), and then connect the axiomatization up to the underlying quantum gravity theory.

In the end, we hope to leave the reader in agreement that constructive axiomatizations in general, and that of EPS in particular, are to be celebrated not just for being testaments to the ingenuity of Reichenbach, Weyl, and EPS, but just as well for the actual service to which they can be put in furthering our understanding of the foundations of spacetime theories.

References

Adlam, Emily. Operational theories as structural realism. *Studies in History and Philosophy of Science*, 94:99–111, 2022. https://doi.org/10.1016/j.shpsa.2022.05.007.

Anandan, Jeeva S. Classical and quantum physical geometry. In *Potentiality, Entanglement and Passion-at-a-Distance*, pages 31–52. Springer, 1997.

Anastopoulos, Charis, Michalis Lagouvardos, and Konstantina Savvidou. Gravitational effects in macroscopic quantum systems: a first-principles analysis. *Classical and Quantum Gravity*, 38(15):155012, 2021.

Anderson, James L. *Principles of Relativity Physics*. Academic Press, 1967.

Andréka, Hajnal, Judit, Madarász István, Németi, Gergely Székely. On Logical Analysis of Relativity Theories. *Hungarian Philosophical Review*, 54:204–222, 2010.

Asenjo, Felipe A. and Sergio A. Hojman. Do electromagnetic waves always propagate along null geodesics? *Classical and Quantum Gravity*, 34(20):205011, 2017.

Audretsch, Jürgen and Claus Lämmerzahl. Establishing the Riemannian structure of space-time by means of light rays and free matter waves. *Journal of Mathematical Physics*, 32(8):2099–2105, 1991.

Audretsch, Jürgen and Claus Lämmerzahl. The conformal structure of space-time in a constructive axiomatics based on elements of quantum mechanics. *General Relativity and Gravitation*, 27(3):233–246, 1995.

Avalos, R., F. Dahia, and C. Romero. A note on the problem of proper time in Weyl space-time. *Foundations of Physics*, 48(2):253–270, 2018.

Bagchi, Arjun, Rudranil Basu, Minhajul Islam, Kedar S. Kolekar, and Aditya Mehra. Galilean gauge theories from null reductions. *Journal of High Energy Physics*, 2022 (4):176, 2022. https://10.1007/JHEP04(2022)176.

Barandes, Jacob A. The stochastic-quantum correspondence. *arXiv preprint arXiv:2302.10778*, 2023a.

Barandes, Jacob A. The stochastic-quantum theorem. *arXiv preprint arXiv:2309.03085*, 2023b.

Barbour, Julian. On general covariance and best matching. In Craig Callender and Nick Huggett, editors, *Physics Meets Philosophy at the Planck Scale*. Cambridge University Press, 01 2001.

Barbour, Julian. Shape dynamics. An introduction, In Felix Finster, Olaf Müller, Marc Nardmann, Jürgen Tolksdorf, Eberhard Zeidler, editors, *Quantum Field Theory and Gravity*. Springer, Basel. 2012. https://doi.org/10.1007/978-3-0348-0043-3_13

Bauer, Andrej. Five stages of accepting constructive mathematics. *Bulletin of the American Mathematical Society*, 54(3):481–498, 2017.

Bekaert, Xavier and Kevin Morand. Connections and dynamical trajectories in generalised Newton-Cartan gravity i: an intrinsic view. *Journal of Mathematical Physics*, 57:022507, 2016.

Bell, John L. and Herbert Korté. Hermann Weyl. In Edward N. Zalta, editor, *The Stanford Encyclopedia of Philosophy*. Metaphysics Research Lab, Stanford University, Winter, 2016.

Belnap, Nuel, Thomas Müller, and Tomasz Placek. *Branching Space-Times: Theory and Applications*. Oxford University Press, 2022.

Belot, Gordon. Why general relativity does need an interpretation. *Philosophy of Science*, 63(S3):S80–S88, 1996. https://10.1086/289939.

Belot, Gordon. *Geometric Possibility*. Oxford University Press, 2011.

Benda, Thomas. A formal construction of the spacetime manifold. *Journal of Philosophical Logic*, 37(5):441–478, 2008. https://10.1007/s10992-007-9075-x.

Benitez, Federico. Selective realism and the framework/interaction distinction: a taxonomy of fundamental physical theories. *Foundations of Physics*, 49(7):700–716, 2019.

Bergmann, P. G. *Introduction to the Theory of Relativity*. Dover, 1942.

Bishop, Errett. *Foundations of Constructive Analysis*. Ishi Press, 1967.

Böhme, Gernot. *Protophysik: Für und wider eine konstruktive Wissenschaftstheorie der Physik*. Suhrkamp, 1976.

Bose, Sougato, Anupam Mazumdar, Gavin Morley, Hendrik Ulbricht, Marko Toros, Mauro Pasternostro, Andrew Geraci, Peter Barker, M. S. Kim, and Gerard Milburn. Spin entanglement witness for quantum gravity. *Physical Review Letters*, 119(24), 2017. https://10.1103/physrevlett.119.240401.

Bridges, Douglas, Erik Palmgren, and Hajime Ishihara. Constructive mathematics. In Edward N. Zalta, editor, *The Stanford Encyclopedia of Philosophy*. Metaphysics Research Lab, Stanford University, Fall, 2022.

Bridgman, P. W. Operational analysis. *Philosophy of Science*, 5(2):114–131, 1938. https://doi.org/10.1086/286496.

Brown, Harvey R. *Physical Relativity: Space-Time Structure from a Dynamical Perspective*. Oxford University Press, 2005.

Brown, Harvey R. and Dennis Lehmkuhl. Einstein, the reality of space and the action–reaction principle. In P. Ghose, editor, *Einstein, Tagore and the Nature of Reality*. Routledge, 2016.

Brown, Harvey R. and Oliver Pooley. The origins of the spacetime metric: Bell's Lorentzian pedagogy and its significance in general relativity. In Craig Callender and Nick Huggett, editors, *Physics Meets Philosophy at the Plank Scale*, pages 256–272. Cambridge University Press, 2001.

Brown, Harvey R. and Oliver Pooley. Minkowski space-time: A Glorious Non-Entity. In Dennis Dieks, editor, *The Ontology of Spacetime*, volume 1 of *Philosophy and Foundations of Physics*, pages 67–89. Elsevier, 2006.

Brown, Harvey R. and James Read. Clarifying possible misconceptions in the foundations of general relativity. *American Journal of Physics*, 84(5):327–334, 2016. https://10.1119/1.4943264.

Brown, Harvey R. and James Read. The dynamical approach to spacetime. In Eleanor Knox and Alastair Wilson, editors, *The Routledge Companion to Philosophy of Physics*. Routledge, 2021.

Bunge, Mario. *Foundations of Physics*. Springer, 1967.

Butterfield, Jeremy. Assessing the Montevideo interpretation of quantum mechanics. *Studies in History and Philosophy of Science Part B: Studies in History and Philosophy of Modern Physics*, 52 (Part A):75–85, 2015. https://10.1016/j.shpsb.2014.04.001.

Butterfield, Jeremy and Henrique Gomes. Functionalism as a species of reduction. In Cristián Soto, editor, *Current Debates in Philosophy of Science: In Honor of Roberto Torretti*, pages 123–200. Springer, 2023.

Cao, ChunJun, Sean M. Carroll, and Spyridon Michalakis. Space from Hilbert space: recovering geometry from bulk entanglement. *Physical. Review. D*, 95:024031, 2017. https://10.1103/PhysRevD.95.024031.

Cao, Tian Yu. Prerequisites for a consistent framework of quantum gravity. *Studies in History and Philosophy of Science Part B: Studies in History and Philosophy of Modern Physics*, 32(2):181–204, 2001. https://doi.org/10.1016/S1355-2198(01)00003-X.

Carnap, Rudolf. *The Logical Structure of the World*. Routledge, 1967.

Carrier, Martin. Constructing or completing physical geometry? On the relation between theory and evidence in accounts of space-time structure. *Philosophy of Science*, 57(3):369–394, 1990. https://doi.org/10.1086/289564.

Carrier, Martin. *Geometric Facts and Geometric Theory: Helmholtz and 20th-century Philosophy of Physical Geometry*, pages 276–292. Akademie, 2018. https://10.1515/9783050070636-016.

Castagnino, M. Some remarks on the Marzke–Wheeler method of measurement. *Il Nuovo Cimento B Series 10*, 54(1):149–150, 1968. https://10.1007/BF02711534.

Chandra, Ramesh. Clocks in Weyl space-time. *General Relativity and Gravitation*, 16(11):1023–1030, 1984.

Chang, Hasok. *Inventing Temperature: Measurement and Scientific Progress: Measurement and Scientific Progress*. Oxford University Press, USA, 2004.

Chang, Hasok. Operationalism. In Edward N. Zalta, editor, *The Stanford Encyclopedia of Philosophy*. Metaphysics Research Lab, Stanford University, Fall, 2009.

Chen, Lu and Tobias Fritz. An algebraic approach to physical fields. *Studies in History and Philosophy of Science Part A*, 89:188–201, 2021. https://doi.org/10.1016/j.shpsa.2021.08.011.

Chen, Lu and James Read. Is the metric signature really electromagnetic in origin? *Philosophy of Physics*, 1, 2023.

Christodoulou, Marios and Carlo Rovelli. On the possibility of laboratory evidence for quantum superposition of geometries. *Physics Letters B*, 792:64–68, 2019. https://doi.org/10.1016/j.physletb.2019.03.015.

Christian Pfeifer. Finsler spacetime geometry in physics. arXiv preprint arXiv:1903.10185, 2019

Cocco, Lorenzo and Joshua Babic. A system of axioms for Minkowski spacetime. *Journal of Philosophical Logic*, 50(1):149–185, 2021.

Coleman, R. A. and H. Korté. Jet bundles and path structures. *Journal of Mathematical Physics*, 21(6):1340–1351, 1980. https://doi.org/10.1063/1.524598.

Coleman, Robert and Herbert Korté. On attempts to rescue the conventionality thesis of distant simultaneity in STR. *Foundations of Physics Letters*, 5(6):535–571, 1992.

Coleman, Robert and Herman Korté. Constructive realism. In *Semantical Aspects of Spacetime Theories*. 1994.

Coleman, Robert Alan and Herbert Korté. Jet bundles and path structures. *Journal of Mathematical Physics*, 21(6):1340–1351, 1980.

Coleman, Robert Alan and Herbert Korté. A new semantics for the epistemology of geometry I: Modeling spacetime structure. In *Reflections on Spacetime*, pages 21–40. Springer, 1995a.

Coleman, Robert Alan and Herbert Korté. A new semantics for the epistemology of geometry II: Epistemological completeness of Newton–Galilei and Einstein–Maxwell theory. In *Reflections on Spacetime*, pages 41–69. Springer, 1995b.

Coleman, Robert Alan and Heinz-Jürgen Schmidt. A geometric formulation of the equivalence principle. *Journal of Mathematical Physics*, 36(3):1328–1346, 1995.

Corry, Leo. *David Hilbert and the Axiomatization of Physics: From* Grundlagen der Geometrie *to* Grundlagen der Physik. Springer, 2004.

Corry, Leo. Hilbert's sixth problem: between the foundations of geometry and the axiomatization of physics. *Philosophical Transactions of the Royal Society A*, 2018.

Covarrubias, G. M. An axiomatization of general relativity. *International Journal of Theoretical Physics*, 32(11):2135–2154, 1993. https://doi.org/10.1007/BF00675025.

Crampin, M. and D. J. Saunders. Projective connections. *Journal of Geometry and Physics*, 57(2):691–727, 2007. https://doi.org/10.1016/j.geomphys.2006.03.007.

Curiel, Erik. General relativity needs no interpretation. *Philosophy of Science*, 76(1):44–72, 2009. https://10.1086/599277.

Curiel, Erik. A Weyl-type theorem for geometrized Newtonian gravity, pages 1–12, 2015. http://arxiv.org/abs/1510.02089.

Curiel, Erik. Kinematics, dynamics, and the structure of physical theory. *arXiv: History and Philosophy of Physics*, 2016.

Curiel, Erik. Schematizing the observer and the epistemic content of theories. *arXiv preprint arXiv:1903.02182*, 2019.

Curry, Sean N. and A. Rod Gover. An introduction to conformal geometry and tractor calculus, with a view to applications in general relativity. In Thierry Daudé, Dietrich Häfner, and Jean-Philippe Nicolas, editors, *Asymptotic Analysis in General Relativity*, London Mathematical Society Lecture Note Series, page 86–170. Cambridge University Press, 2018. https://10.1017/9781108186612.003.

Dautcourt, G. Post-Newtonian extension of the Newton–Cartan theory. *Classical and Quantum Gravity*, 14(1A):A109–A117, 1997. https://10.1088/0264-9381/14/1A/009.

Delhom, Adrià, Iarley P. Lobo, Gonzalo J. Olmo, and Carlos Romero. Conformally invariant proper time with general non-metricity. *European Physical Journal C*, 80(5):415, 2020. https://10.1140/epjc/s10052-020-7974-y.

Deser, Stanley. Gravity from self-interaction redux. *arXiv preprint arXiv:0910.2975*, 2009.

Desloge, Edward A. A simple variation of the Marzke–Wheeler clock. *General Relativity and Gravitation*, 21(7):677–681, 1989. https://doi.org/10.1007/BF00759077.

Dewar, Neil and James Read. Conformal invariance of the Newtonian Weyl tensor. *Foundations of Physics*, 50(11):1418–1425, 2020. https://10.1007/s10701-020-00386-w.

Dewar, Neil, Niels Linnemann, and James Read. The epistemology of spacetime. *Philosophy Compass*, 17(4):e12821, 2022.

Earman, John. Leibnizian space-times and Leibnizian algebras. In *Historical and Philosophical Dimensions of Logic, Methodology and Philosophy of Science*, pages 93–112. Springer, 1977.

Earman, John. Pruning some branches from 'branching spacetimes'. *Philosophy and Foundations of Physics*, 4:187–205, 2008.

Earman, John. Quantum physics in non-separable Hilbert spaces, 2020. http://philsci-archive.pitt.edu/18363/.

Eastwood, Michael. Notes on projective differential geometry. In Michael Eastwood and Willard Miller, editors, *Symmetries and Overdetermined Systems of Partial Differential Equations*, pages 41–60. Springer New York, 2008. https://10.1007/978-0-387-73831-4_3.

Ehlers, Jürgen. Survey of general relativity theory. In Werner Israel, editor, *Relativity, Astrophysics and Cosmology*. Springer, 1973.

Ehlers, Jürgen. On the Newtonian limit of Einstein's theory of gravitation (reprint). *General Relativity and Gravitation*, 51(12):163, 2019. https://10.1007/s10714-019-2624-0.

Ehlers, Jürgen and Robert Geroch. Equation of motion of small bodies in relativity. *Annals of Physics*, 309(1):232–236, 2004. https://doi.org/10.1016/j.aop.2003.08.020.

Ehlers, Jürgen, A. E. Pirani, and Alfred Schild. The geometry of free fall and light propagation (reprint). *General Relativity and Gravitation*, 44(6):1587–1609, 2012 [1972].

Einstein, Albert. On the electrodynamics of moving bodies. *Annalen der Physik*, 17:891–921, 1905. https://10.1002/andp.200590006.

Einstein, Albert. What is the theory of relativity? *The Times*, 1919.

Eppley, Kenneth and Eric Hannah. The necessity of quantizing the gravitational field. *Foundations of Physics*, 7(1–2):51–68, 1977. https://doi.org/10.1007/bf00715241.

Esfeld, Michael. A proposal for a minimalist ontology. *Synthese*, 197(5):1889–1905, 2020.

Esfeld, Michael and Dirk-Andre Deckert. *A Minimalist Ontology of the Natural World*. Routledge, 2017.

Esfeld, Michael and Vincent Lam. Moderate structural realism about space-time. *Synthese*, 160(1):27–46, 2008. https://10.1007/s11229-006-9076-2.

Ewen, Holger and Heinz Jürgen Schmidt. Geometry of free fall and simultaneity. *Journal of Mathematical Physics*, 30(7):1480–1486, 1989. https://10.1063/1.528279.

Feigl, Herbert. Existential hypotheses. realistic versus phenomenalistic interpretations. *Philosophy of Science*, 17(1):35–62, 1950.

Fletcher, Samuel C. Light clocks and the clock hypothesis. *Foundations of Physics*, 43(11):1369–1383, 2013.

Fletcher, Samuel C. On the reduction of general relativity to Newtonian gravitation. *Studies in History and Philosophy of Science Part B: Studies in History and Philosophy of Modern Physics*, 68:1–15, 2019. https://10.1016/j.shpsb.2019.04.005.

Fletcher, Samuel C. Relativistic spacetime structure. In Eleanor Knox and Alastair Wilson, editors, *The Routledge Companion to Philosophy of Physics*. Routledge, 2021.

Folland, Gerald B. Weyl manifolds. *Journal of Differential Geometry.*, 4(2):145–153, 1970. https://10.4310/jdg/1214429379.

Friedman, Michael. *Reconsidering Logical Positivism.* Cambridge University Press, 1999.

Galloway, Gregory J. Null geometry and the Einstein equations. In *The Einstein Equations and the Large Scale Behavior of Gravitational Fields*, pages 379–400. Springer, 2004.

Geroch, Robert and Pong Soo Jang. Motion of a body in general relativity. *Journal of Mathematical Physics*, 16(1):65–67, 1975. https://10.1063/1.522416.

Geroch, Robert and James Owen Weatherall. The motion of small bodies in space-time. *Communications in Mathematical Physics*, 364(2):607–634, 2018. https://10.1007/s00220-018-3268-8.

Giacomini, Flaminia and Časlav Brukner. Quantum superposition of spacetimes obeys Einstein's equivalence principle. *AVS Quantum Science*, 2021. https://api.semanticscholar.org/CorpusID:237416425.

Giacomini, Flaminia, Esteban Castro-Ruiz, and Časlav Brukner. Quantum mechanics and the covariance of physical laws in quantum reference frames. *Nature Communications*, 10, 01 2019. https://10.1038/s41467-018-08155-0.

Giovanelli, Marco. 'But one must not legalize the mentioned sin': phenomenological vs. dynamical treatments of rods and clocks in Einstein's thought. *Studies in History and Philosophy of Science Part B: Studies in History and Philosophy of Modern Physics*, 48(PA):20–44, 2014. http://10.1016/j.shpsb.2014.08.012.

Griffiths, Robert B. The consistent histories approach to quantum mechanics. In Edward N. Zalta, editor, *The Stanford Encyclopedia of Philosophy.* Metaphysics Research Lab, Stanford University, Summer, 2019.

Grimmer, Daniel. The pragmatic QFT measurement problem and the need for a Heisenberg-like cut in QFT. *Synthese*, 202(4):1–45, 2023. https://10.1007/s11229-023-04301-4.

Grinbaum, Alexei. Reconstructing instead of interpreting quantum theory. *Philosophy of Science*, 74(5):761–774, 2007. http://10.1086/525620.

Grinbaum, Alexei. How device-independent approaches change the meaning of physical theory. *Studies in History and Philosophy of Science Part B: Studies in History and Philosophy of Modern Physics*, 58:22–30, May 2017. https://10.1016/j.shpsb.2017.03.003.

Grünbaum, Adolf. *Philosophical Problems of Space and Time.* Dordrecht: Reidel, 1973.

Hansen, Dennis, Jelle Hartong, and Niels A. Obers. Gravity between Newton and Einstein. *International Journal of Modern Physics D*, 28(14):1944010, 2019. https://10.1142/S0218271819440103.

Hansen, Dennis, Jelle Hartong, and Niels A. Obers. Non-relativistic gravity and its coupling to matter. *Journal of High Energy Physics*, 2020 (6):145, 2020. https://10.1007/JHEP06(2020)145.

Harada, Tomohiro, B. J. Carr, and Takahisa Igata. Complete conformal classification of the Friedmann-Lemaître-Robertson-Walker solutions with a linear equation of state. *Classical and Quantum Gravity*, 35(10, April):105011, 2018. https://10.1088/1361-6382/aab99f.

Harada, Tomohiro, Takahisa Igata, Takuma Sato, and Bernard Carr. Complete conformal classification of the Friedmann-Lemaître-Robertson-Walker solutions with a linear equation of state: parallelly propagated curvature singularities for general geodesics. 39 (14, July):145008, 2022. https://10.1088/1361-6382/ac776e.

Hardy, Lucien. The construction interpretation: conceptual roads to quantum gravity, 2018. https://arxiv.org/abs/1807.10980.

Hardy, Lucien. Implementation of the quantum equivalence principle. *Progress and Visions in Quantum Theory in View of Gravity: Bridging foundations of physics and mathematics.* arXiv preprint quant-ph/1903.01289, March 2019. https://link.springer.com/book/10.1007/978-3-030-38941-3.

Harte, Abraham I. Approximate spacetime symmetries and conservation laws. *Classical and Quantum Gravity*, 25(20):205008, 2008.

Hartley, D. Normal frames for non-Riemannian connections. *Classical and Quantum Gravity*, 12, 1995.

Hartong, Jelle, Niels A. Obers, and Gerben Oling. Review on non-relativistic gravity. 12 2022. https://10.3389/fphy.2023.1116888.

Havas, Peter. Simultaneity, conventialism, general covariance, and the special theory of relativity. *General Relativity and Gravitation*, 19(5):435–453, 1987.

Hawking, S. W., A. R. King, and P. J. McCarthy. A new topology for curved space-time which incorporates the causal, differential, and conformal structures. *Journal of Mathematical Physics*, 17(2):174–181, 1976.

Hayashi, Kenji and Takeshi Shirafuji. Spacetime structure explored by elementary particles: microscopic origin for the Riemann-Cartan geometry. *Progress of Theoretical Physics*, 57:302–317, 1977.

Hehl, F. W. and Y. N. Obukhov. *Spacetime Metric from Local and Linear Electrodynamics: A New Axiomatic Scheme*, pages 163–187. Springer, 2006. https://10.1007/3-540-34523-X_7.

Hetzroni, Guy and James Read. How to teach general relativity. *British Journal for the Philosophy of Science*, 2023.

Hilbert, David. *The Foundations of Mathematics.* Harvard University Press, 1927.

Hobson, M. P. and A. N. Lasenby. Weyl gauge theories of gravity do not predict a second clock effect. *Physical Review D*, 102(8):084040, 2020.

Hobson, M. P. and A. N. Lasenby. Note on the absence of the second clock effect in Weyl gauge theories of gravity. *Physical Review D*, 105(2):L021501, 2022.

Hoefer, Carl. The metaphysics of space-time substantivalism. *Journal of Philosophy*, 93(1):5–27, 1996. https://10.2307/2941016.

Hu, B. L. Stochastic gravity. *International Journal of Theoretical Physics*, 38(11):2987–3037, 1999.

Huggett, Nick. The regularity account of relational spacetime. *Mind*, 115(457):41–73, 2006. https://10.1093/mind/fzl041.

Huggett, Nick and Craig Callender. Why quantize gravity (or any other field for that matter)? *Philosophy of Science*, 68(S3):S382–S394, 2001.

Huggett, Nick and Christian Wüthrich. Emergent spacetime and empirical (in)coherence. *Studies in History and Philosophy of Modern Physics*, 44(3):276–285, 2013. https://10.1016/j.shpsb.2012.11.003.

Huggett, Nick and Christian Wüthrich. Out of nowhere: Introduction: The emergence of spacetime. *arXiv preprint arXiv:2101.06955*, 2021.

Huggett, Nick, Niels Linnemann and Mike D., Schneider. *Quantum gravity in a laboratory?*. Cambridge University Press, 2023.

Iliev, B. Z. *Handbook of Normal Frames and Coordinates*. Birkhäuser, 2006.

Inhetveen, R. *Konstruktive Geometrie. Eine formentheoretische Begründung der euklidischen*. Bibliographisches Institut, 1983.

Itin, Yakov and Friedrich W. Hehl. Is the Lorentz signature of the metric of spacetime electromagnetic in origin? *Annals of Physics*, 312(1):60–83, 2004.

Jacobson, Ted. Thermodynamics of spacetime: the Einstein equation of state. *Physical Review Letters*, 75(7):1260, 1995.

Jaksland, R. Entanglement as the world-making relation: distance from entanglement. *Synthese*, 198(10):9661–9693, 2021. https://doi.org/10.1007/s11229-020-02671-7.

Jaksland, Rasmus and Kian Salimkhani. Philosophy, physics, and the problems of spacetime emergence. 2021.

Janich, P. and H. Tetens, editors. *Protophysik Eine Einführung in Protophysik heute*, volume 22. 1985.

Janich, Peter. *Kleine Philosophie der Naturwissenschaften*. C. H. Beck, 1997.

Janich, Peter. *Protophysics of Time: Constructive Foundation and History of Time measurement*, vol. 30. Springer Science & Business Media, 2012.

Janis, Allen. Conventionality of simultaneity. In Edward N. Zalta, editor, *The Stanford Encyclopedia of Philosophy*. Metaphysics Research Lab, Stanford University, Fall, 2018.

Knebelman, M. S. Spaces of relative parallelism. *Annals of Mathematics*, pages 387–399, 1951.

Eleanor Knox. Effective spacetime geometry. *Studies in History and Philosophy of Science Part B: Studies in History and Philosophy of Modern Physics*, 44(3):346–356, 2013.

Köhler, Egon. Measurement of proper time in a Weyl space. *General Relativity and Gravitation*, 9(11):953–959, 1978.

Kretschmann, E. Über den physikalischen Sinn der Relativitätstheorie. *Annalen der Physik*, 53(16):576–614, 1917.

Kriele, M. *Spacetime: Foundations of General Relativity and Differential Geometry*. Lecture Notes in Physics Monographs. Springer, 1999.

Kuchar, Karel. Canonical quantum gravity. *arXiv preprint gr-qc/9304012*, 1993.

Lam, Vincent and Christian Wüthrich. Spacetime is as spacetime does. *Studies in History and Philosophy of Science Part B: Studies in History and Philosophy of Modern Physics*, 64:39–51, 2018.

Lam, Vincent and Christian Wüthrich. Spacetime functionalism from a realist perspective. *Synthese*, 199(2):335–353, 2021.

Lämmerzahl, Claus. Minimal coupling and the equivalence principle in quantum mechanics. *Acta Physica Polonica B*, 29(4):1057–1070, 1998.

Lämmerzahl, Claus and Volker Perlick. Finsler geometry as a model for relativistic gravity. *International Journal of Geometric Methods in Modern Physics*, 15, 02 2018. https://10.1142/S0219887818501669.

Laudan, Larry. *Progress and Its Problems: Towards a Theory of Scientific Growth*, vol. 282. University of California Press, 1978.

Le Bihan, Baptiste and Niels Linnemann. Have we lost spacetime on the way? Narrowing the gap between general relativity and quantum gravity. *Studies in History and Philosophy of Science Part B: Studies in History and Philosophy of Modern Physics*, 65:112–121, 2019. https://doi.org/10.1016/j.shpsb.2018.10.010.

Lee, Jeffrey M. *Manifolds and Differential Geometry*, volume 107. American Mathematical Society, 2022.

Lee, Martin Mok-Don. *Factorization of Multivariate Polynomials*. PhD thesis, Technische Universität Kaiserslautern, 2013.

Lehmkuhl, Dennis. Literal versus careful interpretations of scientific theories: the vacuum approach to the problem of motion in general relativity. *Philosophy of Science*, 84(5):1202–1214, 2017. https://10.1086/694398.

Lehmkuhl, Dennis. The equivalence principle(s). In Eleanor Knox and Alastair Wilson, editors, *The Routledge Companion to Philosophy of Physics*. Routledge, 2021.

Lehmkuhl, Dennis, Gregor Schiemann, and Erhard Scholz. *Towards a Theory of Spacetime Theories*. Birkhauser, 2016.

Leitgeb, Hannes. New life for Carnap's Aufbau? *Synthese*, 180(2):265–299, 2011.

Leitgeb, Hannes. and André Carus. Rudolf Carnap. In Edward N. Zalta, editor, *The Stanford Encyclopedia of Philosophy*. Metaphysics Research Lab, Stanford University, Fall, 2022.

Linnemann, Niels. On the empirical coherence and the spatiotemporal gap problem in quantum gravity: and why functionalism does not (have to) help. *Synthese*, 199(2):395–412, 2021.

Linnemann, Niels and James Read. Comment on 'Do electromagnetic waves always propagate along null geodesics?' *Classical and Quantum Gravity*, 38(23):238001, Dec. 2021a. https://10.1088/1361-6382/ac2c19.

Linnemann, Niels and James Read. On the status of Newtonian gravitational radiation. *Foundations of Physics*, 51(2):53, 2021b. https://10.1007/s10701-021-00453-w.

Linnemann, Niels, James A. M. Read, and Nicholas J Teh. The local validity of special relativity from a scale-relative perspective. *The British Journal for the Philosophy of Science*, 2024. https://doi.org/10.1086/732151.

Linnemann, Niels and Kian Salimkhani. The constructivist's programme and the problem of pregeometry. *arXiv preprint arXiv:2112.09265*, 2021.

Linnemann, Niels and Manus R. Visser. Hints towards the emergent nature of gravity. *Studies in History and Philosophy of Science Part B: Studies in History and Philosophy of Modern Physics*, 64:1–13, 2018.

Linnemann, Niels, Chris Smeenk, and Mark Robert Baker. GR as a classical spin-2 theory? *Philosophy of Science*, 90(5):1363–1373, 2023b.

Livine, Etera R. Quantum uncertainty as an intrinsic clock. *Journal of Physics A: Mathematical and Theoretical*, 56, 2022. https://api.semanticscholar.org/CorpusID:254854470.

Lorenzen, P. *Lehrbuch der konstruktiven Wissenschaftstheorie*. J. B. Metzler, 2016. https://books.google.ca/books?id=_Xy8DQAAQBAJ.

Low, Robert J. Simple connectedness of spacetime in the path topology. *Classical and Quantum Gravity*, 27(10):107001, 2010.

Luc, Joanna. Generalised manifolds as basic objects of general relativity. *Foundations of Physics*, 50(6):621–643, 2020.

Luc, Joanna and Tomasz Placek. Interpreting non-Hausdorff (generalized) manifolds in general relativity. *Philosophy of Science*, 87(1):21–42, 2020.

Majer, Ulrich and Heinz-Jürgen Schmidt. *Semantical Aspects of Spacetime Theories*. BI-Wissenschaftsverlag, 1994.

Malament, David. *Gravity and Spatial Geometry*, volume 114, pages 405–411. Elsevier, 1986. https://doi.org/10.1016/S0049-237X(09)70703-7.

Malament, David B. The class of continuous timelike curves determines the topology of spacetime. *Journal of Mathematical Physics*, 18(7):1399–1404, 1977.

Malament, David B. *Topics in the Foundations of General Relativity and Newtonian Gravitation Theory*. University of Chicago Press, 2012.

Manasse, F. K. and C. W. Misner. Fermi normal coordinates and some basic concepts in differential geometry. *Journal of Mathematical Physics*, 4(6):735–745, 1963. https://10.1063/1.1724316.

Manchak, John Byron. Can we know the global structure of spacetime? *Studies in History and Philosophy of Science Part B: Studies in History and Philosophy of Modern Physics*, 40(1):53–56, 2009. https://doi.org/10.1016/j.shpsb.2008.07.004.

March, Eleanor. Non-relativistic twistor theory: Newtonian limits and gravitational collapse. 2023. https://philsci-archive.pitt.edu/22924/.

March, Eleanor, William J. Wolf, and James Read. On the geometric trinity of gravity, non-relativistic limits, and Maxwell gravitation, *Philosophy of Physics*, 2(1), 15, 2024. https://doi.org/10.31389/pop.80.

Marletto, C. and V. Vedral. Gravitationally induced entanglement between two massive particles is sufficient evidence of quantum effects in gravity. *Physical Review Letters*, 119 (24), 2017. http://10.1103/physrevlett.119.240402.

Marzke, R. F. and J. A. Wheeler. Gravitation as geometry. the geometry of space-time and the geometro-dynamical standard meter. In H. Y. Chiu and W. F. Hoffman, editors, *Gravitation and Relativity*, pages 40–64. W. A. Benjamin, 1964.

James Mattingly. *Is Quantum Gravity Necessary?*. In A. J. Kox and J. Eisenstaedt, editors, *The Universe of General Relativity*. Einstein Studies, vol 11. Birkhäuser, Boston, 2005. doi: 10.1007/0-8176-4454-7_17

Matveev, Vladimir S. and Erhard Scholz. Light cone and Weyl compatibility of conformal and projective structures. *General Relativity and Gravitation*, 52(7):1–9, 2020.

Matveev, Vladimir S. and Andrzej Trautman. A criterion for compatibility of conformal and projective structures. *Communications in Mathematical Physics*, 329(3):821–825, 2014.

Maudlin, Tim. *Philosophy of Physics: Space and Time*. Princeton University Press, 2012.

Menon, Tushar. Algebraic fields and the dynamical approach to physical geometry. *Philosophy of Science*, 86(5):1273–1283, 2019. https://10.1086/705508.

Menon, Tushar, Niels Linnemann, and James Read. Clocks and chronogeometry: rotating spacetimes and the relativistic null hypothesis. *British Journal for the Philosophy of Science*, 71(4):1287–1317, 2018. https://10.1093/bjps/axy055.

Mercati, Flavio. *Shape Dynamics: Relativity and Relationalism*. Oxford University Press, 2018.

Mittelstaedt, Peter. *Rational Reconstructions of Modern Physics*. Springer, 2013.

Mould, Richard A. An axiomatization of general relativity. Proceedings of the American Philosophical Society, 103(3):485–529, 1959. http://www.jstor.org/stable/985480.

Ramírez, Sebastián Murgueitio. On symmetries and springs. *Philosophy of Science*, pages 1–37, 2024. http://10.1017/psa.2023.170.

Nelson, E. *Tensor Analysis*. Princeton University Press, 1967. https://books.google.co.uk/books?id=NgC_QgAACAAJ.

nLab authors. Quantization. http://ncatlab.org/nlab/show/quantization, August 2022. Revision 40.

Norton, John D. General covariance and the foundations of general relativity: eight decades of dispute. *Reports on Progress in Physics*, 56:791–858, 1993.

Norton, John D. Why constructive relativity fails. *British Journal for the Philosophy of Science*, 59(4):821–834, 2008.

Padmanabhan, Thanu. From gravitons to gravity: myths and reality. *International Journal of Modern Physics D*, 17(03n04):367–398, 2008. https://10.1142/s0218271808012085.

Padmanabhan, Thanu. Lessons from classical gravity about the quantum structure of spacetime. *Journal of Physics: Conference Series*, 306:012001, 2011.

Padmanabhan, Thanu. *A dialogue on the nature of gravity*, pages 8–49. Cambridge University Press, 2012. http://10.1017/CBO9780511920998.002.

Penrose, Roger. On gravity's role in quantum state reduction. *General Relativity and Gravitation*, 28(5):581–600, 1996. http://10.1007/BF02105068.

Perlick, Volker. Characterization of standard clocks by means of light rays and freely falling particles. *General Relativity and Gravitation*, 19(11):1059–1073, 1987. http://10.1007/BF00759142.

Perlick, Volker. On the radar method in general-relativistic spacetimes. In *Lasers, Clocks and Drag-Free Control*, pages 131–152. Springer, 2008.

Pfeifer, Christian. Finsler spacetime geometry in physics. *arXiv preprint arXiv:1903.10185*, 2019.

Pfister, Herbert and Markus King. *Inertia and Gravitation: The Fundamental Nature and Structure of Space-Time*. Springer, 2015.

Pincock, Christopher. A reserved reading of Carnap's Aufbau. *Pacific Philosophical Quarterly*, 86(4):518–543, 2005.

Brian-Pitts, J. Space-time philosophy reconstructed via massive Nordström scalar gravities? Laws vs. geometry, conventionality, and underdetermination. *Studies in History and Philosophy of Science Part B - Studies in History and Philosophy of Modern Physics*, 53:73–92, 2016. http://10.1016/j.shpsb.2015.10.003.

Poincaré, Henri. *Science and Hypothesis*. 1905. https://en.wikisource.org/wiki/Science_and_Hypothesis

Pooley, Oliver. Points, particles, and structural realism, December 2005. https://philsci-archive.pitt.edu/2939/. Republished in Dean Rickles, Steven French, and Juha T. Saatsi, editors, *The Structural Foundations of Quantum Gravity*. Oxford University Press, 2007.

Pooley, Oliver. Substantivalist and relationalist approaches to spacetime. In Robert W. Batterman, editor, *The Oxford Handbook of Philosophy of Physics*. Oxford University Press, 2013a.

Pooley, Oliver. Substantivalist and relationalist approaches to spacetime. In Robert Batterman, editor, *The Oxford Handbook of Philosophy of Physics*. Oxford University Press, 2013b.

Pooley, Oliver. Background independence, diffeomorphism invariance, and the meaning of coordinates. In Dennis Lehmkuhl, Gregor Schiemann, and Erhard Scholz, editors, *Towards a Theory of Spacetime Theories*. Birkhäuser, 2017.

Pooley, Oliver and James Read. On the mathematics and metaphysics of the hole argument. *British Journal for the Philosophy of Science*, 0 (ja): null, 2021. https://10.1086/718274.

Putnam, Hilary. *Reason, Truth and History*. Cambridge University Press, 1981.

Read, James. *Explanation, Geometry, and Conspiracy in Relativity Theory*, pages 173–205. Springer International, 2020a. https://10.1007/978-3-030-47782-0_9.

Read, James. Geometrical constructivism and modal relationalism: Further aspects of the dynamical/geometrical debate. *International Studies in the Philosophy of Science*, 33(1):23–41, 2020b. https://10.1080/02698595.2020.1813530.

Read, James. *Special Relativity*. Cambridge University Press, 2023a.

Read, James. *Background Independence in Classical and Quantum Gravity*. Oxford University Press, 2023b.

Read, James and Bryan Cheng. Euclidean spacetime functionalism. *Synthese*, 200(6):466, 2022. https://doi.org/10.1007/s11229-022-03951-0.

Read, James and Thomas Møller-Nielsen. Redundant epistemic symmetries. *Studies in History and Philosophy of Science Part B: Studies in History and Philosophy of Modern Physics*, 70:88–97, 2020.

Read, James and Nicholas J. Teh. The teleparallel equivalent of Newton–Cartan gravity. *Classical and Quantum Gravity*, 35(18):18LT01, 2018. https://10.1088/1361-6382/aad70d.

Read, James, Harvey R. Brown, and Dennis Lehmkuhl. Two miracles of general relativity. *Studies in History and Philosophy of Science Part B: Studies in History and Philosophy of Modern Physics*, 64:14–25, 2018. https://doi.org/10.1016/j.shpsb.2018.03.001.

Reall, Harvey. Part iii: Black hole lectures, 2021. http://www.damtp.cam.ac.uk/user/hsr1000/part3_gr_lectures.pdf. Accessed on 23/12/2021.

Reichenbach, Hans. *The Philosophy of Space and Time*. Dover, 1956.

Reichenbach, Hans. *Axiomatization of the Theory of Relativity*. University of California Press, 1969 [1924].

Robb, Alfred Arthur. *A Theory of Time and Space*. Cambridge University Press, 1914.

Rovelli, Carlo. Ashtekar formulation of general relativity and loop-space nonperturbative quantum gravity: a report. *Classical and Quantum Gravity*, 8(9):1613, 1991.

Rovelli, Carlo. *Quantum Gravity*. Cambridge University Press, 2004.

Rovelli, Carlo and Francesca Vidotto. *Covariant Loop Quantum Gravity: An Elementary Introduction to Quantum Gravity and Spinfoam Theory*. Cambridge University Press, 2015.

Rovelli, Carlo and Francesca Vidotto. Philosophical Foundations of Loop Quantum Gravity. In C. Bambi, L. Modesto, I. Shapiro, editors, *Handbook of Quantum Gravity*. Springer, Singapore, 2024. https://doi.org/10.1007/978-981-99-7681-2_109

Rynasiewicz, Robert. Weyl vs. Reichenbach on Lichtgeometrie. In A. J. Kox and Jean Eisenstaedt, editors, *The Universe of General Relativity*, pages 137–156, Birkhäuser, 2005.

Salimkhani, Kian. The dynamical approach to spin-2 gravity. *Studies in History and Philosophy of Science Part B: Studies in History and Philosophy of Modern Physics*, 72:29–45, 2020.

Salmon, Wesley C. The philosophical significance of the one-way speed of light. *Noûs*, 11(3):253–292, 1977. http://www.jstor.org/stable/2214765.

Sämann, Clemens and Roland Steinbauer. On geodesics in low regularity. *Journal of Physics: Conference Series*, 968(1):012010, 2018. https://10.1088/1742-6596/968/1/012010.

Schelb, Udo. Distinguishability of Weyl- from Lorentz-spacetimes by classical physical means. *General Relativity and Gravitation*, 28(11):1321–1334, 1996a. http://10.1007/BF02109524.

Schelb, Udo. Establishment of the Riemannian structure of space-time by classical means. *International Journal of Theoretical Physics*, 35(8):1767–1788, 1996b.

Schleich, Kristin and Donald M. Witt. Singularities from the topology and differentiable structure of asymptotically flat spacetimes. *arXiv preprint arXiv:1006.2890*, 2010.

Schmidt, Heinz-Jürgen. A minimal interpretation of general relativistic spacetime geometry. In *Reflections on Spacetime*, pages 71–82. Springer, 1995.

Scholz, Erhard. Gauging the spacetime metric—looking back and forth a century later. In *One Hundred Years of Gauge Theory*, pages 25–89. Springer, 2020.

Schröter, Joachim. An axiomatic basis of space-time theory, part i: Construction of a casual space with coordinates. *Reports on Mathematical Physics*, 26(3):303–333, 1988.

Schröter, Joachim. and Udo Schelb. An axiomatic basis of space-time theory, part ii: construction of a C^0-manifold. *Reports on Mathematical Physics*, 31(1):5–27, 1992.

Sklar, Lawrence. Facts, conventions, and assumptions in the theory of space time. *Minnesota Studies in philosophy of Science*, 1977. http://cat.inist.fr/?aModele=afficheN&cpsidt=12734756%5Cnpapers3://publication/uuid/598DB76C-1956-4E70-9422-5208B9F88679.

Spekkens, Robert W. The paradigm of kinematics and dynamics must yield to causal structure. In Anthony Aguirre, Brendan Foster, and Zeeya Merali, editors, *Questioning the Foundations of Physics: Which of Our Fundamental Assumptions Are Wrong?* Springer, 2015.

Stachel, John. Conformal and projective structures in general relativity. *General Relativity and Gravitation*, 43(12):3399–3409, 2011. https://10.1007/s10714-011-1243-1.

Stanford, Kyle. Underdetermination of scientific theory. In Edward N. Zalta, editor, *The Stanford Encyclopedia of Philosophy*. Metaphysics Research Lab, Stanford University, Winter, 2021.

Stein, Howard. Some reflections on the structure of our knowledge in physics. In D. Prawitz, B. Skyrms, and D. Westerståhl, editors, *Logic, Methodology and Philosophy of Science*. Elsevier, 1994.

Synge, John Lighton. A plea for chronometry. *New Scientist*, 19:410–412, 1959.

Synge, John Lighton. *Relativity: The General Theory*. North-Holland, 1960.

Synge, John Lighton. *Relativity: The Special Theory*. North-Holland, 1965.

Tal, Eran. Making time: a study in the epistemology of measurement. *British Journal for the Philosophy of Science*, 67(1):297–335, 2016. https://10.1093/bjps/axu037.

Tamir, Michael. Proving the principle: taking geodesic dynamics too seriously in Einstein's theory. *Studies in History and Philosophy of Science Part B: Studies in History and Philosophy of Modern Physics*, 43(2):137–154, 2012. https://doi.org/10.1016/j.shpsb.2011.12.002.

Teh, Nicholas J. Recovering recovery: on the relationship between gauge symmetry and Trautman recovery. *Philosophy of Science*, 85(2):201–224, 2018. https://10.1086/696375.

Todd, Scott L. and Nicolas C. Menicucci. Sound clocks and sonic relativity. *Foundations of Physics*, 47(10):1267–1293, 2017. https://10.1007/s10701-017-0109-0.

Trautman, Andrzej. Editorial note to: J. Ehlers, F. A. E. Pirani, and A. Schild, The geometry of free fall and light propagation. *General Relativity and Gravitation*, 44(6):1581–1586, 2012.

Tu, Loring W. *An Introduction to Manifolds*. Springer, 2011.

Tumulka, Roderich. Feynman's path integrals and Bohm's particle paths. *European Journal of Physics*, 26(3):L11, 2005.

van Fraassen, Bas C. *The Scientific Image*. Clarendon Press, 1980.

Van Raamsdonk, Mark. Building up space–time with quantum entanglement. *International Journal of Modern Physics D*, 19(14):2429–2435, 2010. https://10.1142/S0218271810018529.

Wallace, David. Who's afraid of coordinate systems? An essay on representation of spacetime structure. *Studies in History and Philosophy of Science Part B: Studies in History and Philosophy of Modern Physics*, 67:125–136, 2019. https://doi.org/10.1016/j.shpsb.2017.07.002.

Wallace, A. Fundamental and emergent geometry in Newtonian physics. *British Journal for the Philosophy of Science*, 71(1):1–32, 2020. https://10.1093/bjps/axx056.

Wang, Zuoqin. Lecture 8: Smooth submanifolds, 2020. http://staff.ustc.edu.cn/wangzuoq/Courses/18F-Manifolds/Notes/Lec08.pdf.

Weatherall, James Owen. Part 1: Theoretical equivalence in physics. *Philosophy Compass*, 14(5):e12592, 2019a. https://doi.org/10.1111/phc3.12592.

Weatherall, James Owen. Part 2: Theoretical equivalence in physics. *Philosophy Compass*, 14(5):e12591, 2019b. https://doi.org/10.1111/phc3.12591.

Weatherall, James Owen. Two dogmas of dynamicism. *Synthese*, 199(S2):253–275, 2020.

Weatherall, James Owen and John Byron Manchak. The geometry of conventionality. *Philosophy of Science*, 81(2):233–247, 2014.

Westman, Hans and Sebastiano Sonego. Events and observables in generally invariant spacetime theories. *Foundations of Physics*, 38(10):908–915, 2008.

Weyl, Hermann. Zur Infinitesimalgeometrie. Einordnung der Projektiven und der Konformen Auffasung. *Nachrichten von der Gesellschaft der Wissenschaften zu Göttingen, Mathematisch-Physikalische Klasse*, 1921:99–112, 1921.

Weyl, Hermann. *Space–time–matter*. Dutton, 1922.

Weyl, Hermann and D. H. Delphenich (trans.). On infinitesimal geometry: relationship with projective and conformal concepts. *Nachrichten von der Gesellschaft der Wissenschaften zu Göttingen, Mathematisch-Physikalische Klasse*, 1921:99–112, 1921. https://www.neo-classical-physics.info/uploads/3/4/3/6/34363841/weyl_-_inf._geom..pdf.

Winnie, John A. The causal theory of space-time. *Minnesota Studies in the Philosophy of Science*, 8:134–205, 1977.

Wolf, William J., James Read, and Quentin Vigneron. The non-relativistic geometric trinity of gravity. *General Relativity and Gravitation*, 56(10):126, 2024a. https://doi.org/10.1007/s10714-024-03308-7

Wolf, William J., Marco Sanchioni, and James Read. Underdetermination in classic and modern tests of general relativity. *European Journal for Philosophy of Science*, 14(4):1–41, 2024b.

Woodhouse, N. M. J. The differentiable and causal structures of space-time. *Journal of Mathematical Physics*, 14(4):495–501, 1973. https://10.1063/1.1666344.

Index

Figures are indicated by an italic *f*.

The manufacturer's authorised representative in the EU for product safety is Oxford University Press España S.A. of el Parque Empresarial San Fernando de Henares, Avenida de Castilla, 2 – 28830 Madrid (www.oup.es/en or product.safety@oup.com). OUP España S.A. also acts as importer into Spain of products made by the manufacturer.

www.ingramcontent.com/pod-product-compliance
Lightning Source LLC
Chambersburg PA
CBHW051400080825
30821CB00006B/37/J

* 9 7 8 0 1 9 8 9 2 2 3 7 7 *